Number Theory and Geometry through History

This is a unique book that teaches mathematics and its history simultaneously. Developed from a course on the history of mathematics, this book is aimed at mathematics teachers who need to learn more about mathematics than its history, and in a way they can communicate it to middle and high school students. The author hopes to overcome, through the teachers using this book, math phobia among these students.

Number Theory and Geometry through History develops an appreciation of mathematics by not only looking at the work of individuals, including Euclid, Euler, Gauss, and more, but also how mathematics developed from ancient civilizations. Brahmins (Hindu priests) devised our current decimal number system now adopted throughout the world. The concept of limit, which is what calculus is all about, was not alien to ancient civilizations as Archimedes used a method similar to the Riemann sums to compute the surface area and volume of the sphere.

No theorem here is cited in a proof that has not been proved earlier in the book. There are some exceptions when it comes to the frontier of current research.

Appreciating mathematics requires more than thoughtlessly reciting first the ten by ten, then twenty by twenty multiplication tables. Many find this approach fails to develop an appreciation for the subject. The author was once one of those students. Here he exposes how he found joy in studying mathematics, and how he developed a lifelong interest in it he hopes to share.

The book is suitable for high school teachers as a textbook for undergraduate students and their instructors. It is a fun text for advanced readership interested in mathematics.

Dr. J. S. Chahal is a professor of mathematics at Brigham Young University. He received his Ph.D. from Johns Hopkins University. After spending a couple of years at the University of Wisconsin as a postdoc, he joined Brigham Young University as an assistant professor and has been there ever since. He specializes in and has published several papers in number theory. For hobbies, he likes to travel and hike. His books, *Fundamentals of Linear Algebra* and *Algebraic Number Theory*, are also published by CRC Press.

Textbooks in Mathematics

Series editors:
Al Boggess, Kenneth H. Rosen

https://www.routledge.com/Textbooks-in-Mathematics/book-series/CANDHTEXBOOMTH

Number Theory and Geometry through History

J. S. Chahal

CRC Press
Taylor & Francis Group
Boca Raton   London   New York

CRC Press is an imprint of the
Taylor & Francis Group, an **informa** business

A CHAPMAN & HALL BOOK

Cover: Public Domain portrait of Euclid by Andre Thevet courtesy of Wikicommons, CC license

First edition published 2025
by CRC Press
2385 Executive Center Drive, Suite 320, Boca Raton, FL 33431, U.S.A.

and by CRC Press
4 Park Square, Milton Park, Abingdon, Oxon, OX14 4RN

CRC Press is an imprint of Taylor & Francis Group, LLC

ISBN: 978-1-041-01175-0 (hbk)
ISBN: 978-1-041-01016-6 (pbk)
ISBN: 978-1-003-61354-1 (ebk)

DOI: 10.1201/9781003613541

Typeset in Nimbus font
by KnowledgeWorks Global Ltd.

Publisher's note: This book has been prepared from camera-ready copy provided by the authors.

To Professors
Takashi Ono
and
Wolfgang M. Schmidt

Contents

IV   Mathematics of the 20th Century

यथा शिखा मयूराणां नागानां मणयो यथा।
तथा वेदाङ्गशास्त्राणां गणितम् मूर्ध्नि स्थितम्॥

As are the crests on the heads of peacocks,
As are the gems on the hoods of cobras,
So is mathematics, at the top of all sciences.

The Yajurveda, circa 600 B.C.

"一年之計，莫如樹穀（谷）；十年之計，莫如樹木；終生之計，莫如樹人。一樹一獲者，穀（谷）也。 一樹十獲者，木也；一樹百獲者，人也。" <管子，權修 第三"

If your plan is for one year, plant rice. If your plan is for ten years, plant trees. If your plan is for one hundred years, educate children.

A Chinese Proverb

Preface

A few years ago, I taught a course on the History of Mathematics at Brigham Young University (BYU) taken mostly by Math Ed majors. These are the students usually in their final year aiming to become school teachers of mathematics. After having taught the course several times, I realized they need to learn more about mathematics than its history, and in a way they can communicate it to the middle and high school students. I think I succeeded in doing so as outlined in this book. I hope the book will not only alleviate the plight of those who suffer from math phobia, but also those who need it. Moreover, every educated person should be familiar with at least a part of this book.

The best place to appreciate wine is to go for wine tasting at its source – the winery. Similarly, the best way to appreciate mathematics is to study the work of those who created it – both known individuals, Euclid, Euler, Gauss, et al. and the ancient civilizations. All civilizations, Babylonian, Chinese, Egyptian, Indian, Native American and even Greek struggled to come up with ways to record a count, such as the number of cattle in a herd or the number of inhabitants in a large metropolis. On this, the final word belongs to some unknown Brahmins (Hindu priests) who devised our current decimal number system that has been adopted now internationally. Still, it is fascinating to look at the failed attempts to understand why they were inadequate to record every number, no matter how large. The same is true about many other aspects of mathematics.

Some people find mathematics boring and intimidating, not because it is so, but because of the way it is taught to them. In India where I was born and had all my education except for my PhD, mathematics was not my cup of tea all the way from primary to middle school. During most of the first 5 years of my primary schooling, all we did during a prescribed period of the day, every day, was thoughtlessly recite, like verses from a holy book, first the 10×10, then 20×20 multiplication tables.

When I moved to the middle school in a nearby larger village, the class sizes went up by ten fold as the students from all nearby smaller villages also had to move there. I was impressed as well as intimidated by some of my classmates who had memorized even the 100×100 multiplication table and could do most calculations in their heads.

However, unlike in America, fortunately at least in my state in India there was not a year of middle or high school when any of the basic subjects – mathematics, languages, science and geography – could be opted out. Not only that, if a student failed the common final exam for all students across the state taking the course on a subject matter, irrespective of who taught it, they had to repeat the whole year before they could move to the next grade. No wonder, some of my classmates were several years older than others. Thus, it was a surprise to me that in America, all students graduate from high school when they are eighteen, like a crop – sowed and harvested on time. Were it like this in India, I would have opted out of math at the earliest opportunity and would have lived with math phobia for the rest of my life.

A few years ago, I sponsored my siblings to migrate to the US. When they arrived in America with their families, among them were two of my nieces, ages eight and sixteen. On their arrival, they enrolled in an elementary and a high school, respectively. To my surprise, they both brought home the same math homework on doing arithmetic with fractions.

No doubt, there are certainly public schools in America that are exceptional. But, in general, especially in poor neighborhoods, their purpose does not seem to educate kids but somehow get them through 12 years of schooling so they can graduate. No wonder their math and geography skills are so poor. In televised political debates, candidates often cannot find on the world map the places where they want to send taxpayers' money.

I am happy that I did not have the option to drop mathematics when I did not want to learn it. Only when I started the ninth grade (the first of a two-year high school), I suddenly discovered that I actually like mathematics. I feel sorry for my classmates who failed to make it to the ninth grade. What inspired me was the proof of the Pythagorean theorem which Euclid composed more than two millennia ago. I found mathematics not only interesting but also beautiful. In American universities, now math majors are required to take a course on "Mathematical Proofs." For me that was it. The art of writing mathematical proofs was already perfected by Euclid.

Before I decided to go for my PhD, I asked one of my professors: how do you do research in mathematics? His answer was brief and to the point. In the world of mathematics, there are lions like Archimedes, Euler, Gauss and Riemann, who kill an animal, eat their lion's share and leave the rest to jackals like me to scratch the bones and finish the job. This book will take the reader to a level to see what exactly he meant. Moreover, the readers can learn, at ease, enough math to scratch some bones themselves.

To be clear, in no way is it a kind of book historians of mathematics write and critique. The issue of credit and ownership of intellectual property is exclusively a Eurocentric concept. In every other civilization, knowledge was a communal property and the credit for creating it went to the whole community.

For example, nobody knows who introduced zero – nothing that is something – to the world. Thus this book is based, in general, more on folklore than quoting the primary, or even secondary sources. It is more interesting to read about the mathematics they created than debate whether it was Leibniz or Newton who "invented" calculus. Although neither of them discovered the Fundamental Theorem of Calculus, both supplied the proof. The concept of limit, which is what calculus is all about, was not alien to ancient civilizations, as we will find out in the first part of this book. Archimedes used a method similar to the Riemann sums to compute the surface area and volume of the sphere.

The book is self-contained in the sense that for the elementary part, no theorem was cited in a proof that has not been proved earlier in the book. There are exceptions when it comes to the frontier of current research. So have fun reading it.

Last but not least, I would like to express my gratitude to all those who offered encouragement, made helpful comments or pointed out errors. In particular, I am thankful to Doug Corey, Barry Mazur, Dipendra Prasad, Kenneth Ribet, Jean-Pierre Serre, Shahid Siddiqui, and Michael Stratoti for taking their time to read the book and getting back to me with their thoughtful suggestions.

Finally, I express my admiration for the impeccable job Lonette Stoddard did in preparing the files for this book.

J. S. Chahal

Advice to the Reader

A substantial part of the book is an easy read requiring no more knowledge than expected of a high school graduate. However, if you are not a professional mathematician or an equivalent of a math major in an American university, you may skip the sections and chapters marked with an asterisk ($*$).

Part I

Arithmetic

1

What Is a Number?

During roughly the period 3000–1500 BC, the world's first four great civilizations flourished along the following river valleys:

The Nile Valley of Egypt,

Mesopotamia, meaning the land between the two rivers (Tigris and Euphrates), also called Babylonia, and now Iraq,

The Indus Valley (Punjab and Sind) of the Indian subcontinent, and

The Huang He Valley of China.

In these agrarian civilizations, a minimal knowledge of mathematics was essential in building waterways, measuring the land and weighing the harvest for tax collection. To anticipate the seasons, it was necessary to establish a calendar based on observing heavenly bodies. An understanding of geometry was needed to build palaces, pyramids, and tombs. Thus it is fairly safe to say that these civilizations had a functional knowledge of the rudimentary concepts of arithmetic, astronomy and geometry. However, as we shall see, even the most advanced civilizations, such as the Greeks, could not come up with a perfect way to represent a number, no matter how large.

These societies had to devise ways to represent (whole) numbers that would be amenable not only for representing but also for calculations. Although the basic idea in every case was that of grouping, some ideas were far superior to others, as can be seen by comparing the Roman numerals to decimal numerals in current use throughout the world. To begin with, it is illuminating to look at this basic notion of grouping.

The concept of numbers one, two, three, and so on, or what we now call natural numbers, must be as old as any human civilization. For example, we would expect that a woman would be able to comprehend how many children she had given birth to. Actually, she must have had a primitive notion of negative numbers as well. For example, if out of seven children she bore, three died, she must have counted the loss of three negatively. The number four of children she was left with is the sum of gain plus loss, in today's notation, $4 = 7 + (-3)$ or simply $4 = 7 - 3$. In fact, the modern definition of $-n$, the negative of a number n, is the number m such that $m + n = 0$. Experiments with birds show that parrots and ravens can compare the number of dots up to six.

One way to define a number is that it is a count. The most natural way to record a count is by bars with one bar for each count. In fact, this way of recording a count is still common in many parts of the world, especially with carpenters and farmers.

When the count is large, it is quite inconvenient to keep track of these bars. To overcome this difficulty, it is common to cross every four marks with a fifth one. For example, the number seventeen can be represented by

The idea is to group the bars into groups of five. Nothing is special about the size of the group. In our example the size of the group called *base* is five, but we could have as easily chosen the base to be six. However, once we pick the base, we must stick with it.

As the count increases and even the number of groups becomes unmanageable, we continue with this idea of "grouping" by organizing groups of groups. So at a glance we can see that Figure 1.1 represents sixty-three:

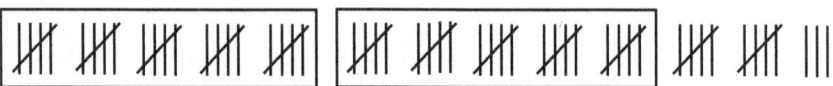

FIGURE 1.1: Idea of grouping.

In this example we have two groups of five fives, two groups of five, and three singles. A shorthand way of writing this tally can be accomplished by representing groups by symbols. Let a single bar | denote a unit, △ denote a group of five, and ▽ denote a group of five fives; i.e., twenty-five. Then our tally would appear (with larger, or more important groups appearing first) as

$$\triangledown\triangledown\triangle\triangle|||$$

Various Numerals to Represent Numbers

Roman Numerals

It is evident that the Romans did not spend much time developing a good system to represent numbers. The ad hoc letters they used for various numbers are illustrated in Figure 1.2. The letters are then combined to express any number up to only a few thousands. To do so, it is recommended that the following rule be used: First write thousands, then hundreds, then tens, and lastly, units. For example, we write 1666 as MDCLXVI.

I	II	III	IV	V	VI	VII	VIII
One	Two	Three	Four	Five	Six	Seven	Eight

IX	X	XI	XII	L	C	D	M
Nine	Ten	Eleven	Twelve	50	100	500	1000

FIGURE 1.2: Roman numerals representing various counts.

One difficulty with Roman numerals is that the representation of a number is not unique. For example, as watchmakers do, four can be written as IIII or IV. Representation can also be ambiguous if one does not know the convention; for example, the number XIX can be interpreted as X(IX), or (XI)X, i.e., nineteen or twenty-one. There are rules for Roman numerals so that parentheses are not needed. If a larger numeral follows a smaller one, these two are taken together. For example, XIX is read as X(IX), not (XI)X. We leave it to the imagination of the reader how to do arithmetic (addition, subtraction, multiplication and division) with Roman numerals.

Egyptian Numerals

A somewhat more sophisticated example of grouping was developed by the Egyptians about 5000 years ago, long before the Romans. In this hieroglyphic system, each of the first several powers of ten is represented by a different symbol as illustrated in Figure 1.3.

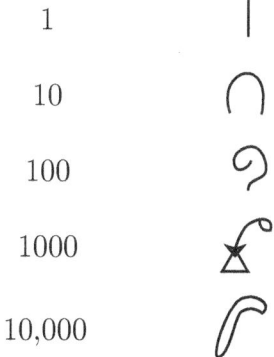

1		
10		
100		
1000		
10,000		

FIGURE 1.3: Egyptian numerals.

Arbitrary whole numbers are represented by appropriate repetition of the symbols. Thus 12,643 can be written as:

Note that unlike our modern place-value system, Egyptians wrote numbers in ascending order of magnitude, beginning with the units on the left. However, the order in which these symbols are arranged is immaterial because there is no place value in this system.

Our knowledge of Egyptian mathematics is based on archeological artifacts such as papyri (see Figure 1.4). The Egyptian way of writing numbers has

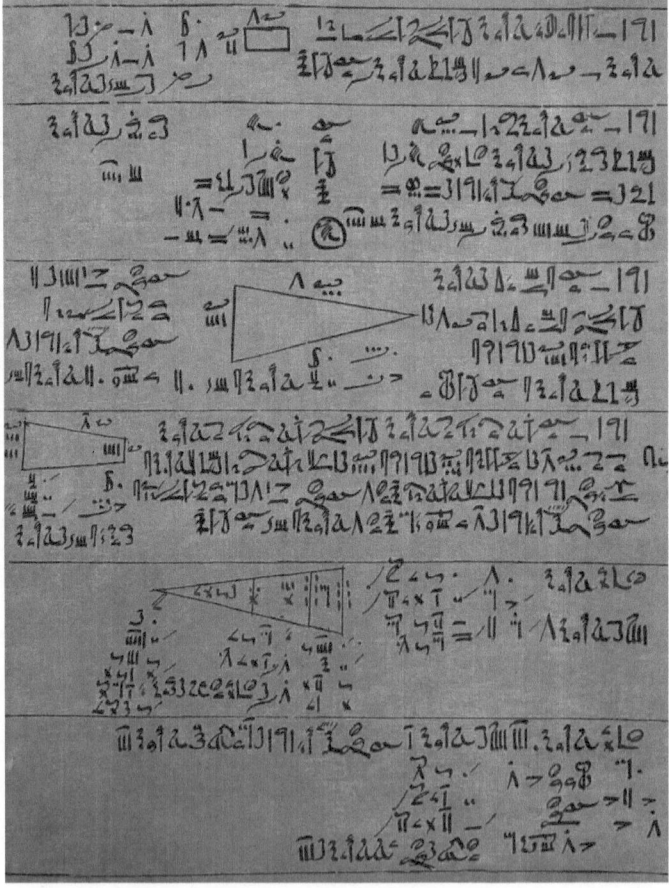

FIGURE 1.4: The Rhind papyrus 1650 BC illustrates the Egyptian method of multiplication.

one major drawback. One needs a symbol for each power of ten. Because the power can be arbitrarily large, there is no end to the need for symbols. For more see [Gil].

Chinese Numerals

The Chinese invented the abacus, a counting board—a table on which counting rods (small bamboo rods about 10 cm long) were manipulated to perform various calculations (see Figure 1.5). Note that the vertical arrangement of the number is listed first followed by the horizontal arrangement.

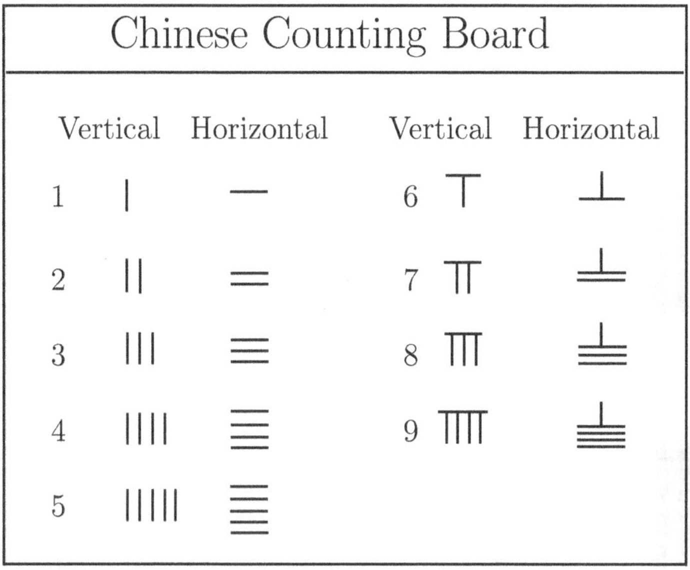

FIGURE 1.5: Chinese counting board with arrangements of the counting rods for the numbers 1 through 9.

To represent numbers greater than ten the rods were set in columns, the rightmost column holding the units. To avoid confusion, the arrangement of rods in neighboring columns alternated between horizontal and vertical. Thus 1156 would be represented as

and 6083 would be represented as

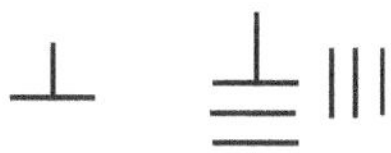

There was no symbol for zero, but a space is left between two non-empty columns.

This system could be confusing, for there is no way of telling if there is indeed a space, were it not for the alternating arrangement of symbols. In the above example, the Chinese, seeing two horizontal arrangements in a row, would immediately know that there was nothing in the hundreds place. But what if there are several consecutive empty places? For example, how can one tell the difference between 530001 and 53000001?

Differing processes were used to perform addition, subtraction, multiplication and division. Negative numbers could be distinguished from positive numbers by coloring. For example, one could color all positive numbers black and all negative numbers red. Manipulations on the counting board were eventually extended to such procedures as solving systems of linear equations and finding numerical solutions to polynomial equations.

Native American Numerals

When the Europeans found the "New World," the land was, according to them, inhabited by primitive and uncivilized people. However, the archaeological discoveries in both South and North America tell a different story. During the "Dark Ages," a description of Europe during approximately the period AD 500–1250, the Americas had some of the most developed civilizations of the time. The most notable are the Mayan and the Inca empires. The Mayan civilization flourished on the Yucatan peninsula of Mexico. During this period, the Europeans were using the Roman numerals, which provide representation of numbers up to only a few thousands and their utility is limited to keeping records involving small numbers, such as that of important dates. On the other hand, the Native Americans had a fairly advanced mathematics for that period in the history of mankind (see [Clo]). The Mayans had a place value system similar to our current decimal number system. Theirs was a vigesimal system (base twenty). They also had a symbol for zero. Their symbols for zero and other digits are shown in Figure 1.6. In this place value system, the numbers were written from top to bottom.

The Mayans were accomplished astronomers. Their year had 365 days, consisting of eighteen monthly periods of twenty days, and an extra period consisting of only five days. Actually, this calendar coexisted with a sacred one used by priests. The priestly calendar had only thirteen vigesimal months. Thus, the sacred calendar year was much shorter with only 260 days in it. The Mayans could predict astronomical events, such as eclipses, for the next several hundred years, with an error of no more than one day. For details, see [Jos, pp. 49–54].

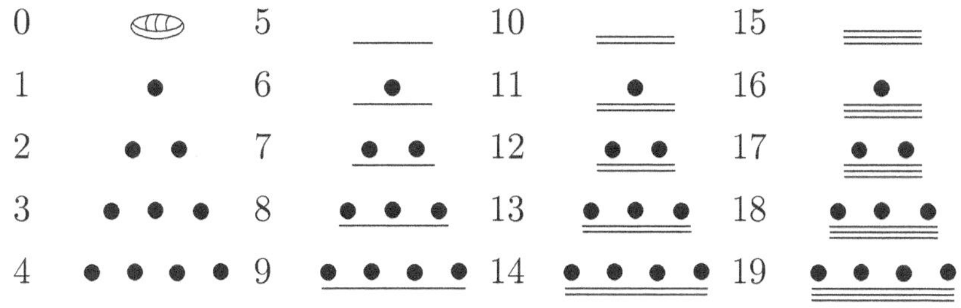

0	⬭	5	—	10	=	15	≡
1	•	6	•̱	11	•=	16	•≡
2	• •	7	•̱•̱	12	• •=	17	• •≡
3	• • •	8	•̱•̱•̱	13	• • •=	18	• • •≡
4	• • • •	9	•̱•̱•̱•̱	14	• • • •=	19	• • • •≡

FIGURE 1.6: Mayan numerals.

Example $\begin{matrix} \bullet \ \bullet \ \bullet \\ \underline{\bullet} \\ \underline{\underline{\ }} \\ ⬭ \\ \underline{\underline{\bullet \ \bullet}} \end{matrix}$ is $\begin{aligned} & 3 \times 20^3 + \\ & 11 \times 20^2 + \\ & 0 \times 20 + \\ & 17 \times 1 = 28,417 \end{aligned}$

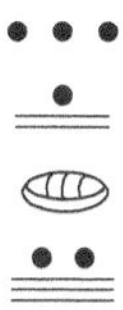

The Incas of South America built a vast empire known as the Kingdom of the Four Directions. A peculiarity with the Inca people is that they had no form of written language. Instead, the Incas communicated throughout the empire through a quipu. A *quipu* is a collection of colored knotted cotton strands (see Figure 1.7). The colors of the strands, their placement on the main strand, the spaces between the strands, and the type and placement of knots on each individual strand all played important parts in interpreting the information.

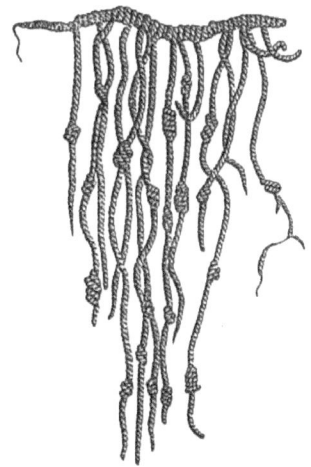

FIGURE 1.7: An Incan quipu.

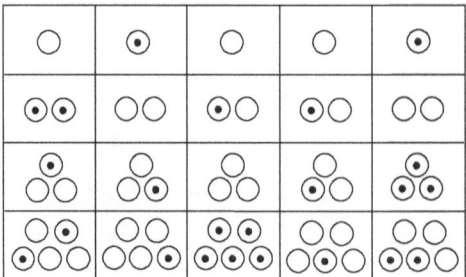

FIGURE 1.8: An Incan yupana.

The color of the strand represented the item that was being counted. For example, yellow would represent gold, red was for the army, and gray was for sheep. If one could not distinguish the objects through color, they were then distinguished through quality. In weaponry, the count of the lance would be first since it was considered the most honorable weapon by the Incas.

The system of counting on the strands was done in base ten. The units were knotted at the bottom of the strand whereas the hundreds or thousands were knotted at the top of the strand. The spacing of the groups of knots had to be precise so that an absence of a group of knots could be noticed when a zero was needed.

Only the *camayoc*, a professional trained in the art of reading and creating quipu, could read the string. The calculations were performed on the Inca abacus, called *yupana* (see Figure 1.8), before they were recorded on a quipu (Figure 1.7). For details see [Jos, pp. 28–41] or [Asc]. The quipucamayoc would create quipu and submit them to the central government so that the ruling class would know the exact economic conditions of all regions of the empire (see Figure 1.9). Thus the Incas could record their history, their laws, and contracts through the quipu without the use of a written language. In a sense, the Incas read a type of machine language common to computers today.

Hindu-Arabic Numerals

Finally, we come to our present and internationally adopted decimal number system, which originated in India. The word decimal is derived from the Sanskrit "dasama" and also perhaps from the Latin "decimus." From India, it spread to Arabia. The Italian mathematician, Fibonacci, who was a frequent visitor to the Islamic world, learned it from Islamic scholars and introduced it to Europe around AD 1220 through his famous book *Liber Abaci* (Book of Calculation). Soon thereafter, the European colonizers, who conquered most of the world, made it international.

An Incan administrator of resources, with the quipu he uses for accounting.

An Incan secretary and accountant who records the dispositions of the royal lords.

FIGURE 1.9: Drawings by Felipe Guaman Poma de Ayala.

Our decimal numerals $1, 2, 3, \ldots$ started in India as

$$-, =, \equiv, \text{Y}, \ldots.$$

Some of these symbols when written hurriedly look like

$$1, 2, 3, 8, \ldots$$

Their evolution to the present form is illustrated in Figure 1.10.

The symbol 0 was invented a little later; first, to indicate an empty place in our place-value number system and then to denote the number zero. Fibonacci begins *Liber Abaci* by introducing these numerals as follows: "The nine figures of the Indians are 9, 8, 7, 6, 5, 4, 3, 2 and 1. With these nine figures, and with the sign 0, which Arabs call zephirum (cipher), any number can be written as we shall show."

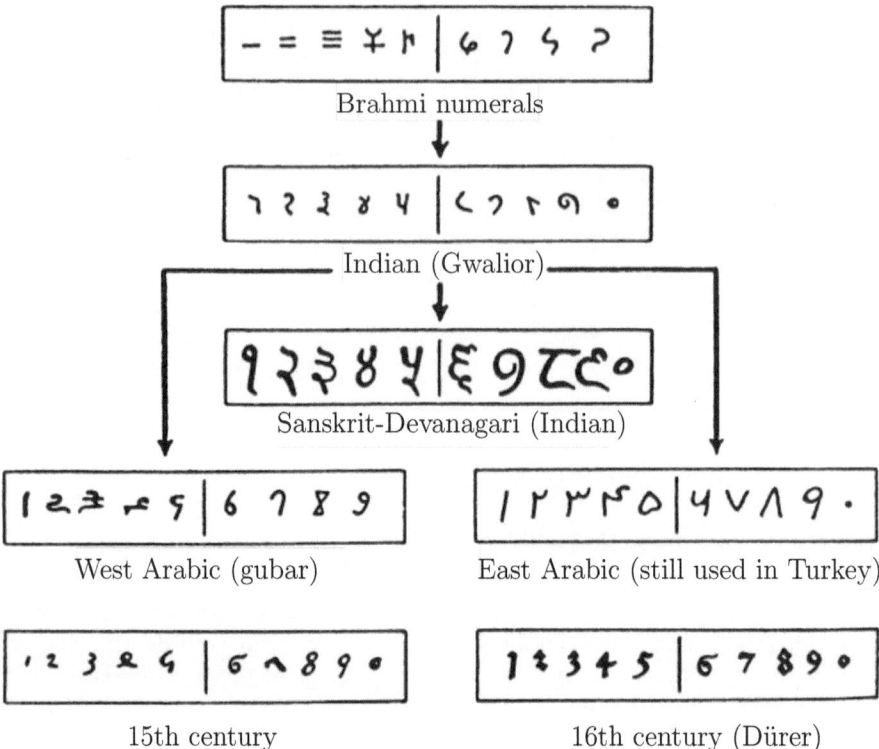

Brahmi numerals

Indian (Gwalior)

Sanskrit-Devanagari (Indian)

West Arabic (gubar)   East Arabic (still used in Turkey)

15th century   16th century (Dürer)

FIGURE 1.10: Development of our modern numerals. (From *Number Words and Number Symbols, A Cultural History of Numbers* by Karl Menninger. Translated by Paul Broneer from the revised German edition. English translation copyright©1969 by The Massachusetts Institute of Technology.)

The contribution of the Hindu-Arabic numerals to the world cannot be overstated. According to the great French mathematician, **Pierre Laplace** (1749–1827) (see [Hog-1, p. 284]):

> India gave us the ingenious method of expressing all the numbers by means of ten symbols, each symbol receiving a value of position as well as an absolute value. It is a profound and important idea which appears so simple to us now that we ignore its true merit. But its very simplicity, the great ease which it has lent to all computations puts our arithmetic in the first rank of useful inventions. We shall appreciate the grandeur of this achievement when we remember that it escaped the genius of Archimedes and Apollonius, two of the greatest men produced by antiquity.

2

Arithmetic in Different Bases

In this chapter, we explain the concept of place value for representing a number. It is so simple, yet so profound and ingenious that according to Laplace, it escaped even the minds of great mathematicians like Archimedes. The speed with which it enables one to perform calculations is astonishing. There is nothing special about base ten in our decimal number system. Any choice of base – the group size – works equally well. The base ten is exceptional because everywhere, people have ten fingers on two hands, which some people still use to count numbers up to ten. In this sense, one may say that it is God given. In South America, the Mayan used base twenty, apparently because we have twenty fingers and toes on our hands and feet.

If we were to pick another base besides ten that would make it easier to work with fractions, then twelve would be a nice pick as one can divide a dozen into halves, thirds, and quarters, something not true with ten. This is the reason clocks will never go metric. The number sixty was a favorite of the Babylonians. They could divide it into two, three, four, five, ten, twelve, fifteen, twenty, and thirty equal parts. The idea of sixty minutes in an hour goes as far back as the Babylonians. From this perspective, a prime base $3, 5, 7, 11, \ldots$ is least desirable.

The Egyptians used base two. The weights found in Harappa and Mohenjo Daro, Indus Valley (see Figure 2.1) suggest that people in this civilization also used base two. The British measurements are a combination of grouping into two and twelve – e.g., pound, shilling, pence for currency and yard, foot, inch for length. The Indian flag has the Ashoka chakra, which is a wheel with twenty-four spokes to represent twenty-four hours in a day, an indication of base three with eight periods in a day. During King Ashoka's reign (273–232 BC), it appeared on his official seal. After the British left, it was adopted as a part of the national emblem of India.

By the 7th century AD, Brahmin scholars had already perfected the place-value decimal number system now used universally. In the previous chapter, we discussed the evolution of the symbols for ten remainders under division by ten. But that is not an issue here. We can use any ten symbols for the ten digits.

FIGURE 2.1: Binary weights from the Indus Valley Civilization (2600–1700 BC).

Now we demonstrate the utility of the place value system of representing numbers in performing arithmetic operations.

If we use base β, we need β digits for zero, one, two, $\ldots$, $\beta - 1$. For example, in base ten, our digits are 0, 1, 2, 3, 4, 5, 6, 7, 8 and 9. For $\beta > 10$, we would have to invent more symbols for extra digits.

A number n may be represented uniquely in any given base. The number nine, which is represented in base ten as 9, has the representation 1001 in base two, i.e., $9 = (1001)_2$. This means that $9 = 1 \cdot 2^3 + 0 \cdot 2^2 + 0 \cdot 2^1 + 1$. The representation of nine in base three is $(100)_3$ for $9 = 1 \cdot 3^2 + 0 \cdot 3 + 0$.

Once we have chosen a particular base, it is as easy to do simple computations such as adding and multiplying in it as in our everyday number system with base ten. Before adding in a different base, let us first examine the process in base ten.

Example 2.1. Adding 369 to 5184 involves the following steps:

1. We first align the numbers so that the digits representing units are in the same column on the right:

$$
\begin{array}{r}
3\ 6\ 9 \\
+\ 5\ 1\ 8\ 4 \\
\hline
\end{array}
$$

2. We consider first the units (i.e., the digits on the far right). Since the sum thirteen of the units is greater than ten (base in which we are working), we place the carryover 1 above the next

column, the column representing groups of ten, and write down the digit 3, leftover from regrouping the units below the summation line.

$$
\begin{array}{r}
1 \\
3\ 6\ 9 \\
+\ 5\ 1\ 8\ 4 \\
\hline
3
\end{array}
$$

3. We do the same again for the column representing tens, and take the carry-over, if any, to the next column representing hundreds. In essence, because we are adding the tens column, we are actually adding 60, 80 and the one ten which we carried over from the units.

$$
\begin{array}{r}
1\ 1 \\
3\ 6\ 9 \\
+\ 5\ 1\ 8\ 4 \\
\hline
5\ 3
\end{array}
$$

4. We continue this process until we reach the last column to get

$$
\begin{array}{r}
1\ 1 \\
3\ 6\ 9 \\
+\ 5\ 1\ 8\ 4 \\
\hline
5\ 5\ 5\ 3
\end{array}
$$

Hence $369 + 5184 = 5553$.

Now we show that it is as easy to do arithmetic in any base as in base ten, the base we are so accustomed to.

Example 2.2. We represent nineteen and fourteen in base three and then add them in a manner similar to that in Example 2.1. In base three,

$$\text{Nineteen is } 2 \cdot 3^2 + 0 \cdot 3 + 1 = (201)_3,$$
$$\text{Fourteen is } 1 \cdot 3^2 + 1 \cdot 3 + 2 = (112)_3.$$

To add 201 and 112 in base three, first note that these representations are not in base ten. So, it is wrong to read them as two hundred one and one twelve. So, read them as two-zero-one and one-one-two in base three.

Note also that we need only the digits $0, 1, 2$. So, discard the digits from 3 to 9. Thus, to add:

1. Align 201 and 112 as shown below with the unit digits in the right most column:

$$
\begin{array}{r}
2\ 0\ 1 \\
+\ \ 1\ 1\ 2 \\
\hline
\end{array}
$$

2. Starting from the units column, we add the digits. Since the sum is three, we carry 1 (group of three) to the next column and place the remainder zero under the column of units as

$$
\begin{array}{r}
1 \\
2\ 0\ 1 \\
+\ 1\ 1\ 2 \\
\hline
0
\end{array}
$$

3. Continuing this process until we reach the last column, we get the sum as shown below.

$$
\begin{array}{r}
1\ \ \ 1 \\
2\ 0\ 1 \\
+\ \ \ 1\ 1\ 2 \\
\hline
1\ 0\ 2\ 0
\end{array}
$$

Thus the sum is $1 \cdot 3^3 + 0 \cdot 3^2 + 2 \cdot 3 + 0$, which is thirty three, as expected.

Let us see this process pictorially, in the spirit of our opening remarks at the beginning of Chapter 1. The nineteen bars and fourteen bars, i.e., $(201)_3$ and $(112)_3$ are represented pictorially as

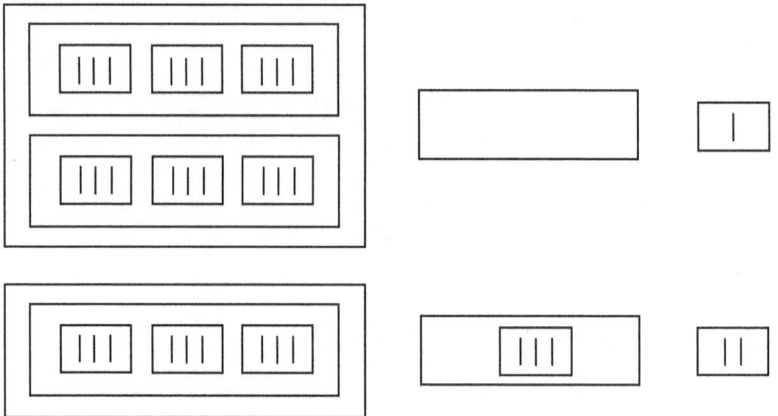

To add, we combine boxes with bars in columns, starting with the right-most column, into larger boxes and move them into the next larger box to get

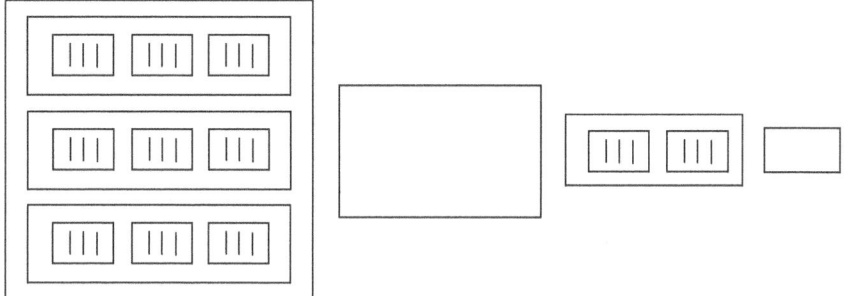

Multiplication

Doing multiplication in other bases is not different from what we are used to in our decimal number system as we now demonstrate. The only difference is to have a multiplication table for the base we are using. We begin with recalling our familiar multiplication table (Figure 2.2).

To find the product 6×7, we look at the entry in the 6th row and 7th column. Because multiplication is commutative, that is $a \times b = b \times a$ (needs justification) this product is the same as the entry in the 7th row and 6th column. Thus we actually need only the part of the table shown in boldface.

Tables for other bases are constructed the same way. For base five, the multiplication table is shown in Figure 2.3. Again, we ignore the entries below the diagonal, since we do not need them.

If the base is two, the table is trivial. To make it look respectable, we also include multiplication by zero (Figure 2.4).

	1	2	3	4	5	6	7	8	9
1	**1**	**2**	**3**	**4**	**5**	**6**	**7**	**8**	**9**
2	2	**4**	**6**	**8**	**10**	**12**	**14**	**16**	**18**
3	3	6	**9**	**12**	**15**	**18**	**21**	**24**	**27**
4	4	8	12	**16**	**20**	**24**	**28**	**32**	**36**
5	5	10	15	20	**25**	**30**	**35**	**40**	**45**
6	6	12	18	24	30	**36**	**42**	**48**	**54**
7	7	14	21	28	35	42	**49**	**56**	**63**
8	8	16	24	32	40	48	56	**64**	**72**
9	9	18	27	36	45	54	63	72	**81**

FIGURE 2.2: Our usual multiplication table.

	1	2	3	4
1	1	2	3	4
2		4	11	13
3			14	22
4				31

FIGURE 2.3: Multiplication tables in base five.

	0	1
0	0	0
1		1

FIGURE 2.4: Multiplication table in base two.

The multiplication table for base twelve is given in Figure 2.5. We need two more symbols to represent digits ten and eleven. In computers, they are A and B. However, conceptually the shape of a digit does not matter. So, to aid memory, we adopt τ for ten and ϵ for eleven. Finally, for the Mayan multiplication table, see Figure 2.6.

Example 2.3. We take base twelve and multiply $4\tau3$ by 1ε.

1. We align these two numbers with the unit digits in the far right column.

$$4\ \tau\ 3$$
$$\times\quad 1\ \varepsilon$$

	1	2	3	4	5	6	7	8	9	τ	ε
1	1	2	3	4	5	6	7	8	9	τ	ε
2		4	6	8	τ	10	12	14	16	18	1τ
3			9	10	13	16	19	20	23	26	29
4				14	18	20	24	28	30	34	38
5					21	26	2ε	34	39	42	47
6						30	36	40	46	50	56
7							41	48	53	5τ	65
8								54	60	68	74
9									69	76	83
τ										84	92
ε											$\tau1$

FIGURE 2.5: Multiplication table in base twelve.

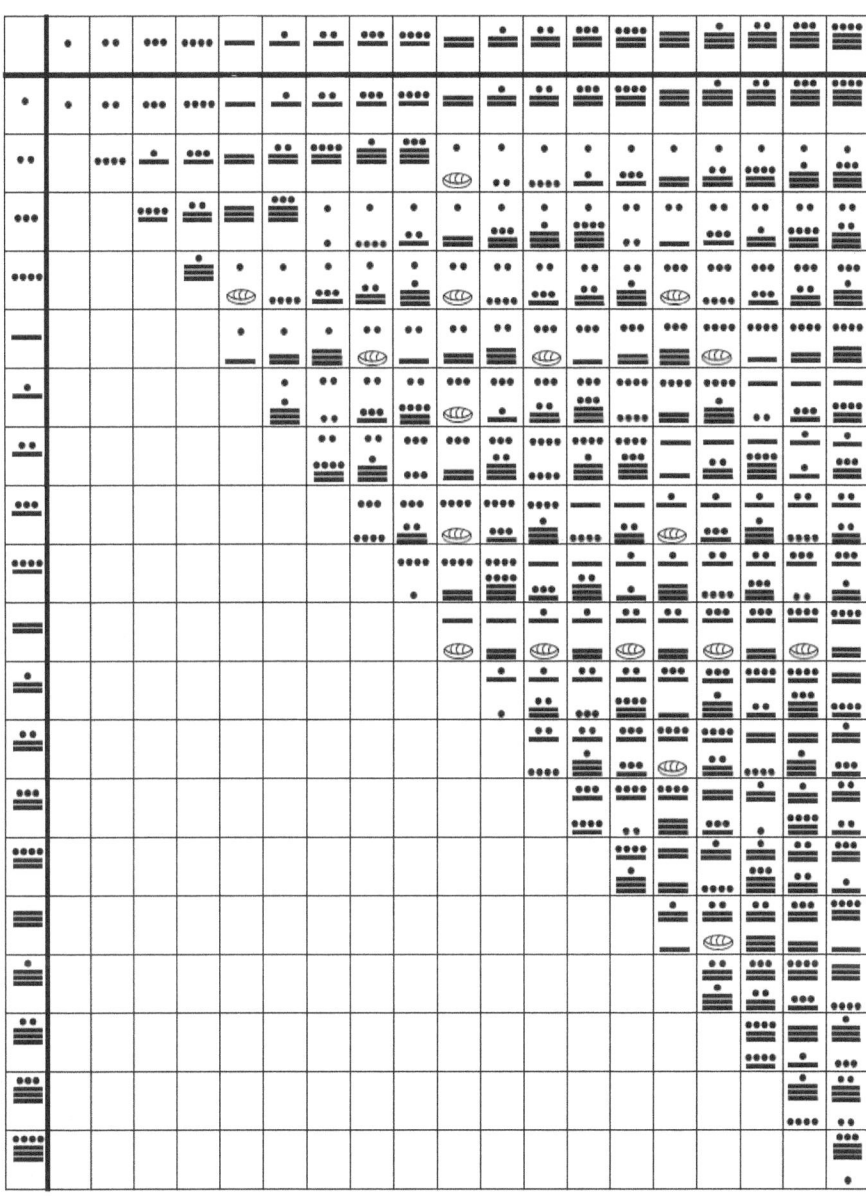

FIGURE 2.6: The Mayan multiplication table – base twenty.

2. Using the table above, we can multiply any two digits. We start at the unit place and find that $3 \times \varepsilon = 29$ (note that it is not twenty-nine, but "two–nine" in base twelve, i.e., two dozens plus nine). We place the units digit 9 under the product bar and carry the digit 2 to the next column as shown below.

$$
\begin{array}{r}
2 \\
4\ \tau\ 3 \\
\times \quad 1\ \varepsilon \\
\hline
9
\end{array}
$$

3. As in our routine multiplication, the digits that are carried over to the next column must be added into the product of digits at the next step. We now multiply τ by ε and add the 2 we carried over from the unit column. From the table, $\tau \times \varepsilon = 92$. Add 2 to get 94. Once again, place the 4 under the product bar and carry over 9. By multiplying 4 by ε, we get 38, but now 9 must be added to the product. We use the procedure outlined above for addition to get $38 + 9 = 45$. Because we have no more digits left to multiply, we place 45 under the product bar as shown below.

$$
\begin{array}{r}
9\ 2 \\
4\ \tau\ 3 \\
\times \quad 1\ \varepsilon \\
\hline
4\ 5\ 4\ 9
\end{array}
$$

4. To begin multiplying with the second digit, we place a zero (0) place holder for obvious reasons under the units column and proceed with the multiplication as shown below.

$$
\begin{array}{r}
9\ 2 \\
4\ \tau\ 3 \\
\times \quad 1\ \varepsilon \\
\hline
4\ 5\ 4\ 9 \\
4\ \tau\ 3\ 0
\end{array}
$$

5. Now the two rows below the product bar are added together as explained earlier to get the final answer.

$$
\begin{array}{r}
4\ 5\ 4\ 9 \\
+\ 4\ \tau\ 3\ 0 \\
\hline
9\ 3\ 7\ 9
\end{array}
$$

Long division is done in a similar manner.

Example 2.4. Take base seven. We want to find the quotient and the remainder when we divide 51643 by 13.

$$
\begin{array}{r}
3460 \\
\hline
13 \enclose{longdiv}{51643} \\
-42 \\
\hline
66 \\
-55 \\
\hline
114 \\
-114 \\
\hline
03 \\
-\ 0 \\
\hline
3
\end{array}
$$

So the quotient $q = 3460$, and the remainder $r = 3$.

Problems for Practice

Perform the indicated arithmetic in the given base. If no base is indicated, it is decimal. In base twelve, τ and ε are digits for ten and eleven.

8734×365

If $5341/17 = q + \dfrac{r}{17}$, find q and r.

$4213 + 1334$ (in base five)

$6132 - 5423$ (in base seven)

$10101 + 11011$ (in base two)

5246×435 (in base seven)

If $4213/21 = q + \dfrac{r}{21}$, find q and r (in base five)

$\varepsilon\tau37 \times 86\tau\varepsilon$ (in base twelve)

Example 2.4 illustrates the Division Algorithm in number theory: When we arrange a count of bars into groups of a given size, the number of groups, called the *quotient*, and the number of leftovers, called the *remainder*, are unique. For example, if the group size is five and the count is seventeen, we have $17 = 3 \times 5 + 2$. Stated mathematically, it goes as below.

Division Algorithm

- *Given natural numbers a and b, we can write*

$$
a = qb + r \quad (0 \le r < b)
$$

with q and b unique.

If a is less than b, the *quotient* q is zero. If the *remainder* r is zero, we say b *divides* a, or b is a *factor* of a. We also say that a is a *multiple* of b.

Now suppose we have fixed the base $b > 1$. Given a natural number n, a count of something, by division algorithm

$$n = q_0 b + d_0 \quad (0 \le d_0 < b).$$

The remainder d_0 represents the *units digit*.

Strictly speaking, the word digit refers to base ten because it is derived from the Greek word for finger.

Repeating the division algorithm with q_0 and b,

$$q_0 = q_1 b + d_1 \quad (0 \le d_1 < b),$$

$$q_1 = q_2 b + d_2 \quad (0 \le d_2 < b).$$

$$\vdots$$

So, $\quad n = q_0 b + d_0$

$$= (q_1 b + d_1)b + d_0$$

$$= (((q_2 b + d_2)b + d_1)b + d_0), \text{ etc.}$$

On clearing the parentheses,

$$n = d_r b^r + d_{r-1} b^{r-1} + \cdots + d_2 b^2 + d_1 b + d_0.$$

This is the *representation of n in base b*.

There is no ambiguity if we write the above expression as $(d_r d_{r-1} \ldots d_2 d_1 d_0)_b$, or simply as $d_r d_{r-1} \ldots d_2 d_1 d_0$, if b is clear from the context. This is the convention we follow because it is universally accepted that b is ten. In computer science, b is two, so that for example,

$$(10001)_2 = 1 \cdot 2^4 + 0 \cdot 2^3 + 0 \cdot 2^2 + 0 \cdot 2 + 1$$

which is seventeen. In computers, the binary representation is used because in an electronic machine, the digits 0 and 1 can be entered as off and on.

Although of no practical value, one can design an optical computer as follows: The sunlight consists of seven fundamental colors, each of the seven digits in base seven can be given a unique fundamental color. If somehow these colors can be entered in a sequence into the machine, we have the "optical" computer, using base seven.

Switching Bases

Our decimal number system is a human invention. The only hurdle to work in another base is that we are so well trained in decimal numerals that it is difficult to think otherwise. As a matter of fact, when we see 52, in say base seven, we are tempted to read it as fifty–two (five tens and two) while it is actually $5 \cdot 7 + 2 =$ thirty–seven. The obstacles are not conceptual, but habitual and linguistic. The names of the numbers themselves are largely decimal-based. The number whose decimal representation is 24, i.e., two tens and four, is twenty–four.

During the transmission of our decimal number system from India to the West, different bases and our terminology got a little bit mixed up.

In English, two peculiar words, *eleven* and *twelve*, seem to be left over from a dozen after making a group of ten. But after twelve, it is thirteen (three and ten), fourteen and so on. Strictly speaking, in our decimal system, eleven and twelve could have been something like oneteen, twoteen. It may be noted that while in English these two words are anomalies, in Indian languages, as well as in Latin, the words for eleven and twelve follow the same pattern as thirteen, fourteen and so on. The Welsh used base nine, hence the word for eighteen (eight and ten) in Welsh is *deu naw* (two nines). Had we used base six, the word for eight would probably sound like twix (two and six). For more, see [Men].

Before working out some examples, the reader is reminded to read a representation, say 63 in base seven, simply as six–three and not sixty–three, for in base seven, 63 is six sevens plus three and not six tens plus three. If a base is not specified, it is understood to be our own decimal base.

Example 2.5. Let $n = 101$, that is, one hundred and one. Suppose we want to express n in binary representation. To find digits, starting with the unit digit, all we need to do is, as explained above, perform long division of $n = 101$ (in decimal system) by base two repeatedly. The remainders are the digits:

$$101 = 50 \cdot 2 + 1$$
$$50 = 25 \cdot 2 + 0$$
$$25 = 12 \cdot 2 + 1$$
$$12 = 6 \cdot 2 + 0$$
$$6 = 3 \cdot 2 + 0$$
$$3 = 1 \cdot 2 + 1$$
$$1 = 0 \cdot 2 + 1.$$

Hence

$$n = (d_r \ldots d_2 d_1 d_0)_2$$
$$= (1100101)_2$$

Example 2.6. Let us find the representation in base seven of the number n whose representation in base five is 41323.

We need to apply the Division Algorithm repeatedly to find the digits. For example, d_0 is given as the remainder of n when divided by seven, that is by 12 (in base five).

But we now have to do the arithmetic in base five. To find the unit digit d_0 by Division Algorithm, we perform the long division (in base five).

$$
\begin{array}{r}
3022 \\
12 \overline{)\ 41323} \\
-\ 41 \\
\hline
03 \\
-00 \\
\hline
32 \\
-24 \\
\hline
33 \\
-24 \\
\hline
4
\end{array}
$$

So, in base five,

$$41323 = 3022 \cdot 12 + 4.$$

Similarly,

$$3022 = 210 \cdot 12 + 2$$
$$210 = 12 \cdot 12 + 11$$
$$12 = 1 \cdot 12 + 0.$$

Hence $n = (41323)_5 = (10624)_7$.

More Practice Problems

We use the notation

$$(d_r \ldots d_1 d_0)_\beta = (\ ?\ )_\gamma$$

to ask what is the representation of the number

$$n = (d_r \ldots d_1 d_0)_\beta$$

in base γ. If β or γ is not specified, it is ten. Use τ and ε as digits for ten and eleven. To check your answers, you must get the same numbers in original bases when you do the problems in reverse.

$36972 = (\ ? \ )_{\text{seven}}$

$2503947 = (\ ? \ )_{\text{twelve}}$

$(251634)_{\text{seven}} = (\ ? \ )_{\text{five}}$

$(2102112)_{\text{three}} = (\ ? \ )_{\text{four}}$

$(1004411)_{\text{five}} = (\ ? \ )_{\text{twelve}}$

3

Arithmetic in Euclid's Elements

The Greek mathematician **Euclid** (3rd century BC) made fundamental contributions to geometry, especially his proof of the Pythagorean theorem in Book I of the *Elements*. Euclid is most famous for his work in geometry. It is said that Greek mathematicians were fond of the slogan, "He who knows not geometry need not enter here."

In this chapter, we shall study his contribution to arithmetic, which are no less fundamental. The truth of the matter is that this is where one still begins their study of number theory. It is not without reason that one of Euclid's theorems on number theory is called the Fundamental Theorem of Arithmetic. Another of his contributions is the Euclidean Algorithm one learns in elementary school to compute the largest common factor of two integers. Once one has learned to work with polynomials, it becomes immediately clear that the Euclidean Algorithm works in exactly the same way to find the greatest common divisor of two polynomials also. As a practical matter, this is how one computes the multiplicative inverse of non-zero elements of finite fields and of the so-called number fields.

Euclid compiled the mathematics of his time into thirteen books, called the *Elements* [Euc]. He devotes Books VII–IX to the study of whole numbers, the subject we call these days number theory, or higher arithmetic. All the numbers considered here are whole numbers, i.e., integers $0, \pm 1, \pm 2, \pm 3, \ldots$. There are some fundamental and self-evident properties of integers that need to be taken for granted. For example, the **Well-ordering Principle**:

- *any non-empty set of positive integers has the least element.*

The central concept in arithmetic is that of divisibility, or what Euclid called measurability, which in modern terminology goes as follows:

Divisibility. A number $b \neq 0$ is said to *divide* a if $a = bc$ for some (whole) number c. We write it as $b|a$. We say that b is a *factor* or *divisor* of a and that a is a *multiple* of b.

The integers 1, 2, 3, 4, 6, 12 are all factors of 12 but none of them is a divisor of 13 as they all leave a non-zero remainder. Thus b divides a if there is no (non-zero) remainder left when a is divided by b (using the Division

Algorithm). If a is a multiple of b, Euclid would say a is measured by b. Every integer is divisible or measured by 1.

Note that it follows at once from the definition that if $b|a$ and $c|b$, then $c|a$. Similarly, if d is a factor of a as well as a factor of b, then d is a factor of $ax + by$ for each pair of integers x and y.

Definition. An integer $p > 1$ is a *prime number* if it has no factors other than ± 1 and $\pm p$.

Thus the prime numbers are

$$2, 3, 5, 7, 11, 13, \ldots.$$

In Euclid's phraseology, "Every number is measured by prime numbers." (See [Euc], Book IX, Prop. 14.) More precisely, we have the following fact.

Fundamental Theorem of Arithmetic. *Every number $n > 1$ is a unique product*

$$n = p_1^{e_1} \cdots p_r^{e_r} \tag{3.1}$$

of powers of distinct primes $p_1, \ldots, p_r$, where $p_1 < \cdots < p_r$.

Euclid gave a proof of this theorem, also called the **Unique Factorization Theorem (UFT)**. However, it was probably Gauss who presented the first correct proof of this important theorem.

Note that we assume that the exponents $e_j \geq 1$. But often it is convenient to allow $e_j = 0$ also.

It is not hard to prove that n is a product of primes. In fact if n is not a prime then

$$n = d_1 d_2 \text{ with } 1 < d_j < n.$$

Now either both d_j are primes, or we can continue factoring them. Since the factors are getting smaller, the process must end.

Uniqueness is more subtle. Uniqueness says that, no matter how we factor say $n = 1728$ (e.g., $1728 = 6 \cdot 9 \cdot 32 = 3 \cdot 2 \cdot 3^2 \cdot 2^5 = 2^6 \cdot 3^3$, or $1728 = 8 \cdot 216 = 8 \cdot 9 \cdot 24 = 8 \cdot 9 \cdot 3 \cdot 8 = 2^6 \cdot 3^3$), at the end we will get $1728 = 2^6 \cdot 3^3$. The prime 2 must occur to power 6 and 3 to power 3, and there is no other way for 1728 to factor into primes.

Proof of Uniqueness. (Gauss) Suppose the theorem is false. We choose the smallest $n > 1$ which has two factorizations, say

$$p_1 \cdots p_r = q_1 \cdots q_s. \tag{3.2}$$

Clearly $p_i \neq q_j$ for every pair i, j of indices, because otherwise we could cancel a common prime factor to get an integer $< n$, also with two factorizations, contradicting the choice of n. We may assume that $p_1 > q_1$. Put $m =$

$$(p_1 - q_1)p_2 \cdots p_r = q_1(q_2 \cdots q_s - p_2 \cdots p_r),$$

by (3.2). Since $m < n$, m has only one factorization. But q_1 is not any of $p_2, \ldots, p_r$. Therefore, q_1 must be a factor of $p_1 - q_1$. If $p_1 - q_1 = q_1 a$, we get $p_1 = q_1(a+1)$. This is a contradiction, because p_1 and q_1 are different primes. Thus the theorem has to be true.   □

The first two of the following corollaries of the Unique Factorization Theorem appear as Propositions 31 and 30, respectively in Book VII of Euclid.

- *Every $n \geq 2$ has a prime factor.*

- *If p divides the product ab, it divides at least one of a and b.*

- *Let n be as in (3.1) above. The exponent of every prime in the unique factorization of n^2 is even.*

The way Euclid states the first assertion, namely, "Every number is measured by a prime" also makes perfect sense, e.g., segments three long to measure a segment twelve long.

The first two assertions follow at once from UFT. The proof of the last assertion is simple algebra:

$$n^2 = (p_1^{e_1} \cdots p_r^{e_r})^2 = p_1^{2e_1} \cdots p_r^{2e_r}.$$

The first question that comes to mind about primes is whether the above list $2, 3, 5, 7, \ldots$ of primes ever ends. The answer is given in Proposition 20, Book IX.

Proposition (Euclid). *There are infinitely many primes.*

Proof. By contradiction: Suppose there are only finitely many primes, say $p_1, \ldots, p_r$. Look at the number

$$N = p_1 \cdots p_r + 1.$$

This number N has a prime factor $p =$ some p_j, by the first corollary above. But p is also a factor of the product $p_1 \cdots p_r$ in which it appears. So, p is also a factor of their linear combination $1 = N - p_1 \cdots p_r$. This is a contradiction, because $p \geq 2$ is not a factor of 1. Hence the assumption that there are only finitely many primes is false.   □

Contradiction in the proof above is not necessary. In fact, a slight modification of the same argument gives a constructive proof. It shows that given any finite set of primes, there is a new prime not in this set.

Two primes p and q $(q > p)$ are a pair of *twin primes* if $q - p = 2$. If one looks at the list of twin primes

$$3, 5; 5, 7; 11, 13; 17, 19; \ldots$$

it seems that this list is also endless. But, in spite of lifelong efforts of many good mathematicians, no one has been able to prove this assertion, called the *twin prime conjecture*.

However, in 2005, Daniel Goldston, János Pintz, and Cem Yildirim [GPY] proved that if the so-called Elliott-Halberstam conjecture is true, then there are infinitely many pairs $p > q$ of primes with $p - q = d$ for some $d \leq 16$. In 2013, Yitang Zhang [Zha] proved, unconditionally, that this is true for a $d \leq 10^7$. In 2014, the bound 10^7 was improved to 246.

Irrational Numbers

The Greeks knew that $\sqrt{2}$ is irrational. However, the Pythagoreans (a close-knit community of mathematicians led by Pythagoras) could not accept the existence of irrational numbers. In fact, Hippasus, an early follower of Pythagoras, was drowned at sea for proving the existence of irrational numbers.

A number (more on this in Chapter 5) α is *rational* if it is a fraction $\alpha = \frac{m}{n}$ of two whole numbers m and n. By changing sign, we can assume that $n > 0$. Some examples of rational numbers are

$$\frac{1}{2}, \frac{3}{5}, \frac{22}{7} \text{ and } 1.414 = \frac{1414}{1000}.$$

A number is an *irrational number* if it cannot be written as a fraction $\frac{m}{n}$. Besides $\sqrt{2}$ there are a lot of other irrational numbers, such as $\sqrt{3}, \sqrt{5}, \sqrt{7}, \ldots$ In fact, we now prove the following assertion.

- *Suppose a is not a perfect square, i.e., a is a positive integer other than $1, 4, 9, 16, 25, \ldots$. Then $\sqrt{a}$ is irrational.*

The proof is again by contradiction: Suppose $\sqrt{a}$ is rational, that is

$$\sqrt{a} = \frac{m}{n}. \tag{3.3}$$

Since a is not a perfect square, the exponent of at least one prime, say p, in the unique factorization of a is odd. By squaring and cross-multiplying, from (3.3) we get

$$an^2 = m^2. \tag{3.4}$$

If we factor $N = m^2$ using the right-hand side (RHS) of (3.4), the exponent of p in its unique factorization is even the last corollary to the UFT while using the left-hand side (LHS), it is odd ($1 +$ an even number). This contradicts the uniqueness of exponent of each prime in the factorization of N. Hence $\sqrt{a}$ cannot be rational.

Since non-squares $2, 3, 5, 6, 7, 8, 10, 11, 12, 13, 14, 15, \ldots$ are for more than squares $1, 4, 9, 16, 25, 36, 49, \ldots$, one can already sense that rationals are far fewer than irrationals. This will be proved in Chapter 5.

Euclidean Algorithm

The *greatest common divisor* (GCD) of two integers a and b, abbreviated as GCD (a, b), is the largest integer which divides both a and b. For example, the greatest common divisor of 30 and 12 is 6, whereas that of 13 and 5 is 1, for there is no integer > 1 that divides both 13 and 5. It is easily verified that the greatest common divisor of a and b is the integer $d \geq 1$ such that

  (i) d divides both a and b, and

  (ii) if another integer c divides both a and b, then c divides d also.

To compute GCD (a, b) it is enough to consider when $a > b > 0$. It follows at once from (i) and (ii) that if

$$a = p_1^{e_1} \cdots p_r^{e_r} \quad (e_j \geq 0)$$

and

$$b = p_1^{f_1} \cdots p_r^{f_r} \quad (f_j \geq 0),$$

then GCD (a, b) is given by

$$\text{GCD}(a, b) = p_1^{\min(e_1, f_1)} \cdots p_r^{\min(e_r, f_r)}.$$

This is, however, the least efficient way to compute GCD (a, b) for it is a hard problem to factor a number into powers of distinct primes. An efficient method to compute GCD (a, b) was devised by Euclid (Book VII, Proposition 2). It is called the Euclidean Algorithm and still remains the most efficient way to compute GCD (a, b).

Euclidean Algorithm. *Write by Division Algorithm*

$$
\begin{aligned}
a &= q_0 b + r_1 && (0 < r_1 < b) \\
b &= q_1 r_1 + r_2 && (0 < r_2 < r_1) \\
r_1 &= q_2 r_2 + r_3 && (0 < r_3 < r_2) \\
&\;\;\vdots \\
r_{j-2} &= q_{j-1} r_{j-1} + r_j && (0 < r_j < r_{j-1})
\end{aligned}
$$

and

$$r_{j-1} = q_j r_j.$$

Then $\text{GCD}(a, b) = r_j$, *the last non-zero remainder in this process.*

Proof. We leave it as an exercise for a motivated reader.   □

Hint: Prove (i) and (ii) above, for which all that is needed is the repeated application of the fact that a common divisor of m and n is also a divisor of $mx + ny$.

Example. Let $a = 13, b = 5$. We write

$$13 = 2 \cdot 5 + 3$$
$$5 = 1 \cdot 3 + 2$$
$$3 = 1 \cdot 2 + 1$$
$$2 = 2 \cdot 1.$$

Hence $\text{GCD}(13, 5) = 1$.

We can use the Euclidean Algorithm to write d as a linear combination

$$ax + by = d,$$

where $d = \text{GCD}(a, b)$.

The Euclidean algorithm works in exactly the same manner for polynomials, in which case, if d is a non-zero constant, by dividing throughout by d, we may assume that $d = 1$.

The following is a standard terminology:

Definition. Two integers (or polynomials) a, b are *coprime*, or *relatively prime*, if $\text{GCD}(a, b) = 1$.

In Chapter 4, we shall see how the existence of a solution of the Diophantine equation,

$$ax + by = 1$$

where a and b are coprime, plays a fundamental role in modern number theory of finite fields (now the backbone of cryptography) and the so-called number fields.

Example. Let $a = 13, b = 5$. As computed above GCD $(a, b) = 1$. Now reversing the steps in the Euclidean Algorithm, we write

$$1 = 3 - 1 \cdot 2$$
$$= 3 - (5 - 1 \cdot 3)$$
$$= 2 \cdot 3 - 5$$
$$= 2(13 - 2 \cdot 5) - 5$$
$$= 2 \cdot 13 + (-5)5.$$

Hence a solution of $13x + 5y = 1 =$ GCD $(13, 5)$ is $x = 2$, $y = -5$.

Exercises.

1. Show that there are arbitrarily large gaps between primes, that is, given $N > 0$, there are primes p and q with

(i) $p - q > N$, and

(ii) there is no prime between q and p.

2. Let GCD $(a, b) = d$. Solve the following linear Diophantine equations $ax + by = d$.

(i) $a = 125$ and $b = 54$

(ii) $a = 13,500$ and $b = 12,709$

(iii) $a = 18,541$ and $b = 12,709$.

3. Given polynomials $a(x) = x^3 - 5x + 1$ and $b(x) = x + 2$, use the Euclidean Algorithm to find polynomials $\lambda(x)$ and $\mu(x)$ so that $\lambda(x)a(x) + \mu(x)b(x) = 1$, the GCD $(a(x), b(x))$.

4

Gauss: Advent of Modern Number Theory

Carl Friedrich Gauss (1777–1856) was born in Braunschweig, Germany and died in Göttingen. Like Wolfgang Amadeus Mozart, he was a child prodigy. He had just entered the primary school when, to keep him busy, his teacher asked him to add all numbers up to n. As n got larger and larger, Gauss found the answer faster and faster. To the curious teacher, he explained his answer as follows: Write the sum s of all numbers up to say 100 in two ways, forward and backward, and add

$$s = 1 + 2 + 3 + \quad \cdots \quad + 99 + 100$$
$$s = 100 + 99 + 98 \quad\quad + 2 + 1$$
$$\overline{2s = 101 + 101 + 101 \quad \cdots \quad + 101 + 101}$$

so, $s = \frac{100 \cdot 101}{2}$.

In college courses on mathematical proofs by induction, the proof of the formula

$$1 + 2 + 3 + \cdots + n = \frac{n(n+1)}{2}$$

is a favorite example. However, by the age of 21, when many college students are still struggling to graduate, Gauss has already finished writing *Disquisitiones Arithmeticae* [Gau], which he published at the age of 24 in 1801 and moved onto the study of astronomy, physics, and other branches of mathematics; although he famously said,

Mathematics is the queen of all sciences and number theory is the queen of mathematics.

In mathematics, many theorems are named after people who do not have much to do with them. For example, the only thing John Pell (1611–1685) had to do with the equation named after him was to ask **Leonhard Euler** (1707–1783) a question about it. For the sake of reference, Euler named it the Pell equation. Pythagoras neither proved the theorem named after him nor was the first to discover it. It was Euclid who gave a complete proof of the Pythagorean theorem, which by the way should more appropriately be called the *Fundamental Theorem of Geometry*. Euclid, whom many consider to be the father of modern mathematics, was the first to record in the *Elements*, the *Fundamental Theorem of Arithmetic* [Euc, XI, Prop. 14]. However, it was

Gauss who gave its first correct proof. In his doctoral thesis, Gauss also proved the *Fundamental Theorem of Algebra*, which states that *a polynomial of degree $n \geq 1$ has exactly n roots, counted properly*. With first correct proofs of two of the four fundamental theorems of mathematics, anyone would have already earned a glorious place in the history of mathematics. But for Gauss, it was just the beginning.

Number Theory of Gauss

Gauss created a brand new arithmetic, called the *modular arithmetic*. To "mod out" is a part of our daily life. To find the day of the week, we mod out by seven. In this process, all that matters are the remainders zero to six. All multiples of seven are Sundays, multiples of seven plus one are Mondays, etc. To find the fractional part of a number, we mod it out by 1. The value of a trig function, say $\sin(2\pi t)$, depends only on "$t \bmod 1$", i.e., the fractional part of t.

Gauss went a step further. He showed how to add and multiply two remainders to get another remainder making the set of remainders a number system in its own right. Before explaining it, let us be clear that in this context, by "numbers," we mean the integers $0, \pm 1, \pm 2, \ldots$.

Choose any number m larger than one, Gauss called *modulus*. For the sake of exposition, let us stick with $m = 7$, the number of days in a week. By the division algorithm, a number n can be expressed uniquely (i.e., in one and only one way) as $n = 7q + r$ with remainder r from 0 to 6. When we drop the largest multiple $7q$ of 7, the remainder r is called the *reduction of n* (mod 7). The set of all numbers $7q + r$ is the *residue class* (mod 7) of r and is denoted by $\bar{r}$. There are exactly seven residue classes (mod 7), namely, $\bar{0}, \bar{1}, \ldots, \bar{6}$. In the context of the days of the week, $\bar{0}$ represents all Sundays, current, past, and future, $\bar{1}$ all Mondays, and so on.

According to Gauss, the residue classes (mod 7) are "added" and "multiplied" by the following rule:

$$\bar{r} \oplus \bar{s} = \bar{t}$$

$$\bar{r} \odot \bar{s} = \bar{u}$$

with t the reduction of $r + s$ (mod 7) and u the reduction of $r \cdot s$ (mod 7). Note the slight difference in notation. By $a \equiv b \pmod{m}$, Gauss means a and b are in the same residue class (mod m). By $a = r \pmod{m}$, we mean r is the smallest non-negative representative of the residue class (mod m) to which a belongs. For example, $\bar{3} \oplus \bar{4} = \bar{0}$ because $3 + 4 = 7 = 0 \pmod 7$. Similarly, $\bar{3} \odot \bar{5} = \bar{1}$, because $3 \cdot 5 = 15 = 1 \pmod 7$.

For the sake of convenience, from now on the circles around $+$ and dot, as well as bars over r, s, t, u will be invisible, e.g., we write $3 + 4 \equiv 0 \pmod 7$ and $3 \cdot 5 \equiv 1 \pmod 7$ simply as $3 + 4 = 0$ and $3 \cdot 5 = 1$, keeping in mind that the modulus is 7.

Recall the definition of the *negative* $-a$ of a number a. It is the number b such that $a + b = 0$. Therefore, since in the set $= \{0, 1, \ldots, 6\}$ of remainders, $3 + 4 = 0$, $-4 = 3$. Similarly, the *reciprocal* or *inverse* of a non-zero a is b with $a \cdot b = 1$. Since $3 \cdot 5 = 1$, so $\frac{1}{3} = 5$.

With this new arithmetic, the set $\{0, 1, \ldots, 6\}$ of seven remainders, denoted by $\mathbb{F}_7$, becomes the so-called *field of seven elements*. Within $\mathbb{F}_7$, this arithmetic follows all the rules of our usual arithmetic. Moreover, we can write and solve equations within $\mathbb{F}_7$, as Gauss did. We do it in the following fun game, which illustrates how cryptography works.

Cryptography

Cryptography is not a new science. Julius Caesar used to send encrypted messages to his generals by moving each letter of the alphabet three places to the right in a circular way. For example, "attack at midnight" was sent as "dwwdfn dw plgqljkw." His generals knew how to recover the message.

Suppose you want to send to a friend by email the last four digits 3162 of your credit card number in a disguised form, so that if it falls into the wrong hands, you are safe. How will you do it? All your digits are in $\mathbb{F}_7 = \{0, 1, \ldots, 6\}$. If they are not, choose a higher modulus. The following is an idea to scramble (or encrypt) them: Use the equation (key)

$$\boxed{s = 3r + 4.}$$

Thus we convert the digits 3162 as follows:

$$r = 3 \rightarrow s = 3 \cdot 3 + 4 = 13 = 6 \pmod 7$$

$$1 \rightarrow 3 \cdot 1 + 4 = 0$$

$$6 \rightarrow 3 \cdot 6 + 4 = 1$$

$$2 \rightarrow 3 \cdot 2 + 4 = 3.$$

Therefore, the four digits are encrypted as 6013. Assume you shared the key with your friend who solves the equation for r in terms of s, i.e.,

$$r = \frac{s-4}{3} = \frac{1}{3}s + \frac{1}{3}(-4)$$

$$= 5s + 5 \cdot 3 = 5s + 1.$$

So the reverse key is

$$\boxed{r = 5s + 1.}$$

To recover the original digits from 6013, your friend uses this reverse key:

$$s = 6 \rightarrow r = 5 \cdot 6 + 1 = 3$$

$$0 \rightarrow 5 \cdot 0 + 1 = 1$$

$$1 \rightarrow 5 \cdot 1 + 1 = 6$$

$$3 \rightarrow 5 \cdot 3 + 1 = 2.$$

You see that decryption takes 6013 back to 3162.

In real life, communication involves dozens of characters (letters, digits, punctuation marks, including the blank space, etc.), say all in all 47. Now choose the modulus $m = 47$ and assign the 47 elements $\{0, 1, \ldots, 46\}$ of $\mathbb{F}_{47}$ in some order to your 47 characters. Encrypt/decrypt them in a message with $m = 47$ instead of $m = 7$ as in our example.

There are ways to *break the code* (guess the key) by analyzing the text if the key is too simple as in our example. So, companies like banks that transfer huge amounts of data every day, hire mathematicians to make the key more and more complicated to stay ahead in the game. More on this in Chapter 13.

Complex Numbers

In many ways, polynomials

$$a + bx + cx^2 + dx^3 + \cdots$$

behave like integers $0, \pm 1, \pm 2, \ldots$. In particular, both have the Division Algorithm. Given polynomials $f(x)$ and $m(x)$ with $\deg m(x) \geq 1$, we can perform long division to get the quotient $q(x)$ and remainder $r(x)$:

$$f(x) = q(x)m(x) + r(x)$$

with $\deg r(x) < \deg m(x)$.

Thus we can do modular arithmetic with polynomials mod $m(x)$ exactly as we did it with integers.

If we choose modulus $m(x) = 1 + x^2$, the reduction $r(x) \bmod m(x)$ of $f(x)$ is a polynomial of degree at most one, so $r(x) = a + bx$. To do the modular arithmetic $(\bmod m(x))$ with polynomials, all multiples of $1 + x^2$, in particular $1 + x^2$ itself, are 0. If we denote $\bar{x}$, the residue class of $x \bmod (1 + x^2)$, by i we have $1 + i^2 = 0$, i.e., $i = \sqrt{-1}$, and the residue classes mod $(1 + x^2)$ are $a + bi$ which are exactly the complex numbers. Thus, the modular arithmetic mod $(1 + x^2)$ is the same as that of complex numbers.

It is plausible and worth a research topic that this way of constructing complex numbers was Gauss's motivation for introducing modular arithmetic.

Note that like integers $m = 7$ and 47, the polynomial $m(x) = 1 + x^2$ also does not factor non-trivially. If for integers, we chose the modulus $m = 10$ (or for polynomials the modulus $m(x) = 1 - x^2$), the modular arithmetic is more subtle. For example, $(\bmod 10)$ there is no a in the set $\{0, 1, \ldots, 9\}$ representing the residue classes $(\bmod 10)$ with $4a = 1 \pmod{10}$ because no multiple of 4 under division by 10 will leave remainder 1. Thus $\frac{1}{4}$ does not exist.

Suppose $\bmod m$ (or $\bmod m(x)$ for polynomials), the inverse exists. One should wonder how to find it. If m is small, say $m = 7$, one can find it by trying all non-zero remainders 1–6. But what if m is huge, e.g., has one million digits? The answer was already known to Euclid more than two millennia ago as discussed in Chapter 3.

Application of Number Theory: Construction of Septadecagon

In the *Elements*, Euclid showed how to construct (by which it is always meant using compass and straightedge only) an equilateral triangle (Proposition 1 of Book I, or simply Prop. I.1), a square (Prop. I.46). But one of the most interesting propositions (Prop. IV.11) in the *Elements* is the construction of a regular pentagon.

By a regular n-gon $(n > 2)$, we mean an n-sided geometric figure in the plane with all sides of equal length. For example, a 3-gon is the equilateral triangle, a 4-gon is the square, 5-gon is the pentagon, and the 17-gon is the septadecagon.

Constructing an n-gon, dividing the circle into n equal parts, constructing the angle $\frac{2\pi}{n}$ and constructing a line segment of length $\cos\frac{2\pi}{n}$ are all essentially the same construction.

To obtain a $2n$-gon from an n-gon is a trivial task. So, assume that n is odd.

Suppose p, q $(p \neq q)$ are odd prime numbers. By the Euclidean Algorithm, one can write $1 = aq + bp$ for integers a and b (Chapter 3). Therefore, multiplying throughout by $\frac{2\pi}{pq}$,

$$\frac{2\pi}{pq} = a\frac{2\pi}{p} + b\frac{2\pi}{q}.$$

Thus, if we can construct angles $\frac{2\pi}{p}$ and $\frac{2\pi}{q}$, we can also construct angle $\frac{2\pi}{pq}$.

The Euler ϕ-function $\phi(n)$ counts the number of integers a, $0 < a < n$, such that a and n have no common factor larger than 1. For example, $\phi(6) = 2$, because only 1 and 5 have no common factor with 6 larger than 1.

It can be proved that for the n-gon to be constructible, it is necessary that the Euler ϕ-function $\phi(n)$ is a power of 2. But, for $n = p^d$ ($d > 1$, p an odd prime), $\phi(n)$ is divisible by p. (See Chapter 13.) Therefore, the question reduces to: for which odd primes p, a p-gon is constructible?

For two millennia, it had been asserted that the only n-gon for n odd that can be constructed are for $n = 3, 5$ and 15. However, on March 29, 1796, at age 18, Gauss discovered that it is possible to construct the 17-gon, an event that motivated his decision to become a professional mathematician.

How Did Gauss Do It?

To construct line segments of newer and newer lengths, one intersects lines and circles. Algebraically, this amounts to obtaining solutions

$$x = \frac{-b \pm \sqrt{b^2 - 4ac}}{2a}$$

of quadratic equations

$$ax^2 + bx + c = 0,$$

repeatedly. To construct a 17-gon, Gauss showed ([Gau, §365]) that $\cos \frac{2\pi}{17}$ is a solution of successive quadratic equations. More precisely, he showed that

$$\cos \frac{2\pi}{17} = \frac{1}{16}\left[-1 + \sqrt{17} + \sqrt{34 - 2\sqrt{17}} \right.$$

$$\left. + 2\sqrt{17 + 3\sqrt{17} - \sqrt{34 - 2\sqrt{17}} - 2\sqrt{34 + 2\sqrt{17}}} \right].$$

Five years later, in the last chapter of *Disquisitiones Arithmeticae*, titled "Equations Defining Division of Circle into Equal Parts", Gauss gave a complete answer to which n-gons are constructible.

- *For an* n-gon *with* n *odd to be constructible, it is necessary and sufficient that* n *is a product*

$$n = p_1 \ldots p_r$$

of distinct Fermat primes $p_1, \ldots, p_r$.

Fermat primes are the primes of the form $2^{2^m} + 1$ with $m \geq 0$. First few Fermat primes are $3, 5, 17, 257$, and 65537, for $m = 0, 1, 2, 3$ and 4.

Pierre de Fermat (1607–1665), considered to be the Father of Modern Number Theory, had conjectured around 1647 that all numbers of the form $F_m = 2^{2^m} + 1$ are primes. However, in 1732, Euler showed that $F_5 = 4,294,967,297$ is divisible by 641, thus not a prime. Actually, up to now, the only known Fermat primes are the ones which were already known to Fermat. An unsolved problem in number theory is to find a new Fermat prime or prove there is no other Fermat prime. So, there are only 31 odd-gons known to be constructible by ruler and compass.

In §365 of *Disquisitiones Arithmeticae*, Gauss recounted:

It is certainly astonishing that although the geometric divisibility of the circle into three and five was already known in Euclid's time, nothing was added to this discovery for 2000 years. And all geometers had asserted that, except for these sections and the ones that derive directly from them, there are no others that can be effected by geometric constructions.

The quote shows how much importance Gauss attached to his discovery of the 17-gon, and he had every reason to be proud of his work on n-gons. He opened an entirely new era in number theory. For two millennia, the arithmetic and geometry that Euclid left behind seemed totally unrelated. It was his study of the n-gons that for the first time, Gauss brought together arithmetic, algebra, and geometry in a way never thought possible before, and the depth of his work is breathtaking. In order to prove his assertion of constructibility of n-gons, he initiated the study of the so-called cyclotomic fields, now a part of Galois theory.

Proud of his 17-gon discovery, Gauss requested that a 17-gon be inscribed on his tombstone. Due to some difficult circumstances, this wish was never fulfilled. However, in Braunschweig, where Gauss was born, there is a statue of him standing on a 17-pointed star.

Equations over Finite Fields*

In the section on cryptography, we already dealt with equations over finite fields, a subject initiated by Gauss. Although Gauss did not use our current terminology of groups, rings, and fields, he was certainly doing the same thing in his own way. He also initiated counting the number of solutions of equations over finite fields. This culminated not only in P. Deligne's proof of the Weil conjecture, for which in 1978, he was awarded the Fields Medal (an equivalent of the Nobel Prize in mathematics) but also in the creation of the modern (scheme theoretic) version of algebraic geometry by **Alexander Grothendieck** (1928–2014), primarily to prove the Weil conjecture.

To explain what all this means, suppose $m > 1$ is an arbitrary but fixed modulus. If $f(x)$ is a polynomial with integer coefficients and for an integer r, $f(r)$ is divisible by m, Gauss called r a solution of the congruence

$$f(x) \equiv 0 \pmod{m}.$$

If we replace each coefficient of $f(x)$ by its reduction mod m, the polynomial so obtained is called the *reduction* of $f(x) \bmod m$. By the so-called Chinese Remainder Theorem (Chapter 14), there is no loss of generality in assuming that $m = p^d$ is a prime power. In particular, if $d = 1$, solving the congruence

$$f(x) \equiv 0 \pmod{p}$$

is the same as solving its reduction $\pmod{p}$

$$f(x) = 0,$$

which we call an equation over $\mathbb{F}_p$, with solutions x in $\mathbb{F}_p$.

The simplest such equations in one and two variables are the linear equations

$$by = c$$

and

$$ax + by = c.$$

The first one has a unique solution $y = \frac{c}{b}$. The second one has exactly p solutions, $x = \frac{c-by}{a}$, one for each y in $\mathbb{F}_p$.

As in our usual arithmetic, the quadratic equation

$$Ax^2 + By + C = 0 \tag{4.1}$$

has solutions in $\mathbb{F}_p$ $(p > 2)$, namely,

$$x = \frac{-B \pm \sqrt{B^2 - 4AC}}{2A} \tag{4.2}$$

provided for $B^2 - 4AC = a \bmod p$, the equation

$$x^2 = a \tag{4.3}$$

has a solution in $\mathbb{F}_p$. If it doesn't, there is no point in trying to find it. In order to determine the existence of a solution of $x^2 = a$ in $\mathbb{F}_p$, Gauss proved the Law of Quadratic Reciprocity.

Note that for $p = 2$, the denominator $2A$ in (4.2) is zero, which makes it invalid.

Gauss called a in equation (4.3) a *quadratic residue* (mod p) if the equation (4.3) has a solution in $\mathbb{F}_p$, otherwise a *quadratic non-residue* (mod p). A few years earlier, Legendre had already discovered the *Law of Quadratic Reciprocity* but gave an erroneous proof of it. Gauss not only gave a correct proof but gave over half a dozen of them. Since then, countless proofs of the Quadratic Reciprocity have appeared, not all of them correct.

The *Legendre symbol*

$$\left(\frac{a}{p}\right) = \begin{cases} 1 & \text{if } a \text{ is a quadratic residue mod } p \\ -1 & \text{otherwise} \end{cases}$$

satisfied the rule

$$\left(\frac{a\,b}{p}\right) = \left(\frac{a}{p}\right)\left(\frac{b}{p}\right). \tag{4.4}$$

Thus, to study equation (4.3), it suffices to assume that a is a prime other than p.

If p is reasonably small, one can solve equation (4.3) by inspection. However, if p is exceedingly large, say has a million digits, that is not possible. In this case, we use the Law of Quadratic Reciprocity to reduce p to a manageable size.

If $p > q > 2$ are two primes, we can consider q as an element of $\mathbb{F}_p$ and the reduction of p (mod q) that of $\mathbb{F}_q$.

Law of Quadratic Reciprocity*

• *If p, q ($p \neq q$) are odd primes,*

$$\left(\frac{q}{p}\right) = \left(\frac{p}{q}\right)(-1)^{\frac{p-1}{2} \cdot \frac{q-1}{2}}. \tag{4.5}$$

Example. Let us try to solve

$$x^2 = 15$$

in $\mathbb{F}_{101}$. For that, we need to determine whether the Legendre symbol $\left(\frac{15}{100}\right)$ is 1 or -1. By (4.4),

$$\left(\frac{15}{101}\right) = \left(\frac{3 \cdot 5}{101}\right) = \left(\frac{3}{101}\right)\left(\frac{5}{101}\right).$$

By the Law of Quadratic Reciprocity,

$$\left(\frac{3}{101}\right) = \left(\frac{101}{3}\right)(-1)^{\frac{100}{2} \cdot \frac{2}{2}}$$

$$= \left(\frac{2}{3}\right) = -1,$$

because 2 is not a square in $\mathbb{F}_3$.

Similarly,

$$\left(\frac{5}{101}\right) = \left(\frac{101}{5}\right)(-1)^{\frac{100}{2} \cdot \frac{4}{2}}$$

$$= \left(\frac{1}{5}\right) = 1$$

because 1 is a square in every $\mathbb{F}_p$. Therefore, $\left(\frac{15}{101}\right) = -1$, showing that $\sqrt{15}$ does not exist in $\mathbb{F}_{101}$.

Without this law, we had to square all numbers from 1 to 100, reduce them mod 101 to see if 15 is among them, not an easy job even for a fairly small $p = 101$.

Cubic Equations*

In §358 of *Disquisitiones Arithmeticae*, Gauss studied the cubic equation

$$x^3 + y^3 = 1 \tag{4.6}$$

over $\mathbb{F}_p$. Its significance was not realized until 1921 when **Emil Artin** (1898–1962) discovered the so-called Riemann hypothesis for elliptic curves over finite fields.

Gauss counted the number N_p of solutions of (4.6) with x, y in $\mathbb{F}_p$ for various primes p. He proved the following:

- *If $p \equiv 1 \pmod 3$, $N_p = p$.*

Otherwise, $4p = a^2 + 27b^2$ for some integers a, b. If the sign of a is so chosen that $a \equiv 1 \pmod 3$, then $N_p = p + a$. In particular,

$$|a| = |p - N_p| \le 2\sqrt{p}. \tag{4.7}$$

This is the conjecture Artin stated in his thesis for a larger class of equations, which came to be known as the Riemann hypothesis for elliptic curves over finite fields (next section). First note that the substitution

$$x = \frac{6}{X} + \frac{Y}{6X}, \quad y = \frac{6}{X} - \frac{Y}{6X}$$

transforms equation (4.6) to

$$Y^2 = X^3 - 432 \tag{4.8}$$

whereas

$$X = \frac{12}{x + y}, \quad Y = 36 \frac{x - y}{x + y}$$

takes (4.8) back to (4.6). This sets up a one-to-one correspondence between the solutions of (4.6) and that of (4.8). Therefore, the count N_p remains unchanged. We must take $p \ne 2, 3$ because in both $\mathbb{F}_2$ and $\mathbb{F}_3$, 6, 12, 36, and 432 are all zero, so both of these substitutions as well as equation (4.8) break down.

Equation (4.8) is a special case of the equation

$$E : y^2 = x^3 + ax + b \tag{4.9}$$

over $\mathbb{F}_p$ with $4a^3 + 27b^2 \ne 0$ in $\mathbb{F}_p$. Suppose N_p denote again the number of solutions of (4.9) with x, y in $\mathbb{F}_p$.

E. Artin conjectured in his 1924 thesis the following estimate for N_p called the

- *Riemann hypothesis for elliptic curves over finite fields:*

$$|p - N_p| \le 2\sqrt{p}. \tag{4.10}$$

Gauss had already stated and proved the germinal case of (4.10).

Riemann Hypothesis*

One may wonder why the inequality (4.10) is called the Riemann hypothesis. For this, Artin attached an analytic object to the elliptic curve E defined by

(4.9), its zeta function $Z_E(u)$ with $u = p^{-s}$ ($s = \sigma + it$), which turns out to be a rational function of u:

$$Z_E(u) = \frac{1 - a_p u + pu^2}{(1 - u)(1 - pu)},$$

with $a_p = p - N_p$. The complex number $u = p^{-s}$ is a zero of $Z_E(u)$ if and only if p^s satisfies

$$f(t) = t^2 - a_p t + p = 0.$$

Now $|a_p| \le 2\sqrt{p}$ if and only if the two roots of $f(t)$ given by

$$t = \frac{a_p \pm \sqrt{a_p^2 - 4p}}{2}$$

are complex conjugates $\alpha, \bar{\alpha}$ (by Vieta's theorem $\alpha + \bar{\alpha} = a_p$, $\alpha\bar{\alpha} = p$). This is so if and only if $\alpha\bar{\alpha} = p \Leftrightarrow |\bar{\alpha}| = |\alpha| = |p^s| = p^\sigma = \sqrt{p} \Leftrightarrow \sigma = \frac{1}{2}$.

See Chapter 15 for the original Riemann hypothesis that Riemann made in 1859.

The inequality (4.10), also called the *Hasse inequality*, was proved by **Helmut Hasse** (1898–1979) in 1933.

If

$$f(x, y) = 0$$

is an equation of an "irreducible" and "smooth" curve of degree $d \ge 1$, its genus is the non-negative integer $g = \frac{(d-1)(d-2)}{2}$. For example, lines and conics have genus zero, whereas the elliptic curve defined by (4.9) has genus $g = 1$.

André Weil (1906–1998) proved the Hasse inequality for every curve with $g \ge 1$ as

$$|a_p| = |p - N_p| \le 2g\sqrt{p}. \tag{4.11}$$

Note that (4.10) is a special case of (4.11) with $g = 1$.

Weil's proof is exceedingly difficult. Simpler proofs were given by Bombieri [Bom] and Schmidt [Sch-2].

An equation

$$f(x, y, z) = 0$$

in three variables with coefficients in $\mathbb{F}_p$ defines a *surface* and

$$f(x_1, \ldots, x_n) = 0,$$

in n variables is a *hypersurface*. An intersection of several hypersurfaces is called a *variety*. After proving the estimate (4.10) for curves, A. Weil also made a sweeping conjecture, called the *Weil conjecture*, about the number of solutions of a variety in all finite fields that contain $\mathbb{F}_p$ as a subfield. The

Weil conjecture was proved by P. Deligne for which he was awarded the Fields Medal in 1978.

Thus, the seminal case of the Weil conjecture had already been stated and proved by Gauss in *Disquisitiones Arithmeticae.*

There is a lot more to Gauss's work in number theory than can be covered in this chapter, such as his work on quadratic forms. But, it suffices to say that even this much should be more than enough to convince the reader that Gauss is indeed the greatest number theorist of all time.

5

Numbers Beyond Rationals

Arithmetic of Rational Numbers

A *rational number*, or simply a *rational*, is a partition of an integer m into $n \geq 1$ equal parts, which is written as $\frac{m}{n}$. If the absolute value of m is less than n, we call $\frac{m}{n}$ a *fraction*. If $m > n > 1$ and $m = qn + r$ $(0 < r < n)$, it is customary to write $\frac{m}{n} = q + \frac{r}{n}$ as $q\frac{r}{n}$, e.g., $\frac{17}{5} = 3\frac{2}{5}$. But, we will not do this. One can see that for $d \geq 1$, $\frac{m}{n} = \frac{dm}{dn}$. Thus, we can assume that the GCD $(m, n) = 1$. On the other hand,

$$\frac{a}{b} + \frac{c}{d} = \frac{ad}{bd} + \frac{bc}{bd} = \frac{ad + bc}{bd}.$$

Similarly, $\frac{a}{b} \cdot \frac{c}{d} = \frac{ac}{bd}$. Since $\frac{a}{b} \cdot \frac{b}{a} = \frac{ab}{ba} = 1$ (a, b both nonzero), $\frac{1}{\frac{a}{b}} = \frac{b}{a}$. As a matter of notation, we write $\underbrace{a \cdots a}_{m-\text{times}} = a^m$ and $\frac{1}{a} = a^{-1}$, etc.

Before saying a number is irrational, one must be clear what is meant by a "number." Only then it makes sense to ask whether a "given number" is rational or not. Moreover, one must know how to add, subtract, multiply, and divide the numbers under consideration. The precise way of doing this was not achieved until recently by **Augustin-Louis Cauchy** (1789–1857) and a little later by **Richard Dedekind** (1831–1916), using the so-called Cauchy sequences and Dedekind cuts, respectively.

Taking more and more decimal digits for, say π, is a Cauchy sequence that represents π. On the other hand, Dedekind cuts defined in a later section is a slick way to fill the gaps that correspond to irrational numbers.

Real Numbers

So, what do we mean by "numbers" when we claim a "given number" is irrational? The most down-to-earth way of perceiving them is as distances of points on the *real line* from a chosen origin, marked zero and a point marked

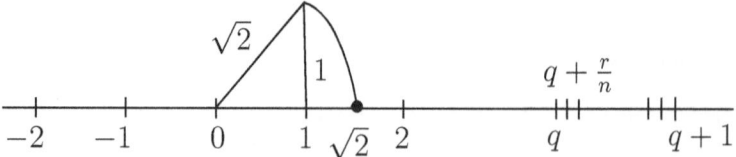

FIGURE 5.1: The real line.

1 to represent the unit distance. The negatives are on the other side of zero (see Figure 5.1).

By definition, the *real numbers*, or simply the *reals*, are points on the real line. There is an obvious way to mark on it the rational point $\frac{m}{n} = q + \frac{r}{n}$ (see Figure 5.1). As we have seen in Chapter 3, $\sqrt{2}$ is irrational, so there is a gap at a distance $\sqrt{2}$ from 0, left after marking all rationals on the real line. The points on the real line that do not correspond to rationals are *irrationals*.

The addition and subtraction of reals as points on the real line is obvious but not their multiplication and division when both are irrational. For example, what do we mean by $\sqrt{13}/\pi$, i.e., to divide $\sqrt{13}$ into π equal parts? Equivalently, what is the point on the real line at a distance $\sqrt{13}/\pi$ units from the origin?

We can certainly construct a line segment of length $\sqrt{13}/\pi$ using the proportionality of the side lengths of similar triangles (see Figure 5.2).

However, to state the proportionality of the side lengths of similar triangles, we are already assuming what the ratio $\sqrt{13}/\pi$ means. Thus, we are in a situation analogous to egg and chicken, which we will deal with in Chapter 6.

To get out of the circular argument, suppose x, y are two real numbers, i.e., points on the real line. Let $x_1, x_2, x_3, \ldots$ be a sequence of rational numbers getting closer and closer to x, and similarly $y_1, y_2, y_3, \ldots$ for y. Then, xy is the points to which the sequence $x_1 y_1, x_2 y_2, x_3 y_3, \ldots$ of rationals gets closer and

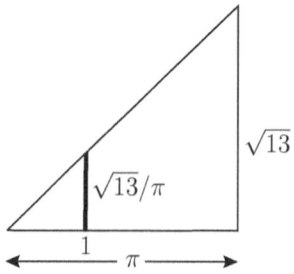

FIGURE 5.2: Quotient of irrational numbers.

closer. To avoid having to prove that such a point exists, we identify real numbers with the so-called Cauchy sequences of rational numbers. For example, we can say that π is the Cauchy sequence $3, 3.1, 3.14, 3.141, 3.1412, \ldots$ of rational numbers $3, \frac{31}{10}, \frac{314}{100}, \frac{3141}{1000}, \frac{31412}{10000}, \ldots$ Two Cauchy sequences $x_1, x_2, x_3, \ldots$ and $y_1, y_2, y_3, \ldots$ of rational numbers such that $x_n - y_n$ approaches zero are deemed to be the same, or equivalent.

A slick way to define reals which fills the gaps belonging to the irrational numbers on the real line was conceived by Dedekind, discussed in the next section.

Dedekind's Construction of Real Numbers

Dedekind, a student of Gauss, revolutionized algebra by generalizing the concept of natural and prime numbers to ideals and prime ideals. Unlike numbers, ideals are sets (of numbers). He then proved that integral ideals in the so-called Dedekind domains factor uniquely into powers of distinct prime ideals, just like naturals numbers factor uniquely into products of powers of distinct primes.

Using the same idea – thinking of a number as the set it represents – Dedekind gave the most slick construction of real numbers. He defined a cut (on the real line), now called in his honor a *Dedekind cut*, α as a partition $A \mid B$ of the rational numbers into two disjoint non-empty sets A and B, such that all numbers in A are less than every number in B, and A contains no greatest number. The set B may or may not have a smallest number in it. If B has a smallest number, the cut corresponds to that rational number. Otherwise, that cut defines a unique irrational number which, loosely speaking, fills the "gap" between A and B. We will define a real number to be a Dedekind cut α. Moreover, if we allow A or B to be empty, we can add $\pm\infty$ as well to the reals.

Example. The sets A and B of rationals $r < 2$ and $r \geq 2$, respectively, define the cut that corresponds to the rational number 2, whereas the sets A and B of rationals r satisfying $r^2 < 2$ and $r^2 \geq 2$ define the irrational number $\sqrt{2}$.

Dedekind cuts can be added and multiplied in an obvious way. If A_1, A_2 are two sets of rationals, their sum $A_1 + A_2$ is the set of all rational numbers $r_1 + r_2$ with r_1 in A_1 and r_2 in A_2. Similarly, their product $A_1 A_2$ is the set of all rational numbers $r_1 r_2$ with r_1 in A_1 and r_2 in A_2.

Now if $\alpha_1 = A_1 \mid B_1$ and $\alpha_2 = A_2 \mid B_2$ are two Dedekind cuts, their sum and product are the cuts $\alpha_1 + \alpha_2 = A_1 + A_2 \mid B_1 + B_2$ and $\alpha_1 \alpha_2 = A_1 A_2 \mid B_1 B_2$. The other operations on cuts can be defined in a similar manner.

Complex Numbers

Some of the problems mathematicians had been occupied with since antiquity are solving equations of small degrees up to 5, in particular quadratic and cubic equations $ax^2 + bx + c = 0$ and $ax^3 + bx + c = 0$. However, even the simple quadratic equation $x^2 + 1 = 0$ has no solutions in real numbers. So, the "imaginary" numbers were introduced in order for these equations to have solutions in them. The quadratic formula states that the two solutions of $ax^2 + bx + c = 0$ are

$$x = \frac{-b \pm \sqrt{b^2 - 4ac}}{2a}.$$

Put $D = b^2 - 4ac$. Then both the solutions are real numbers if and only if $D \geq 0$. For $D < 0$, let $|D| = -D$. Then the above formula is

$$x = \frac{-b \pm \sqrt{|D|}\sqrt{-1}}{2a}.$$

So, we let i denote $\sqrt{-1}$, and call a number $z = a + ib$ with a, b real, a *complex number*. It is clear that we can solve any quadratic equation if we can solve $x^2 + 1 = 0$. To do this we "mod out" polynomials in x by $1 + x^2$, an idea introduced by Gauss, as explained in Chapter 4.

Abraham de Moivre (1667–1754) noted that some trig identities could be re-expressed simply by the following well-known formula which bears his name, *de Moivre's formula*:

$$(\cos \theta + i \sin \theta)^n = \cos n\theta + i \sin n\theta.$$

It was Leonhard Euler who denoted $\sqrt{-1}$ by i and defined the complex exponentiation e^z for $z = x + iy$. Clearly, x may be assumed to be zero (because of the desired property $e^{z_1+z_2} = e^{z_1}e^{z_2}$). Euler's formula

$$e^{iy} = \cos y + i \sin y$$

is actually the definition of complex exponentiation. In the special case $z = i\pi$, this ties the five most important numbers $0, 1, e, i, \pi$ in the equation

$$e^{i\pi} + 1 = 0$$

called the *Euler identity* which the Nobel laureate physicist Richard Feynman called "the most remarkable formula in mathematics."

As mentioned earlier, one of the most fundamental contributions to the theory of complex numbers was made by Gauss in his doctoral thesis. In an equivalent form, it is the Fundamental Theorem of Algebra which says:

- *Every polynomial of positive degree has a complex root.*

Historical Remark

The other two symbols e, π were also used for the first time in mathematics by Euler.

The idea of a complex number as a point in the plane, now called the *complex plane*, was first described by Danish-Norwegian mathematician Caspar Wessel (1745–1818) in 1799, although it had been anticipated as early as 1685 by John Wallis (1616–1703). It was Gauss who called the number system in which $\sqrt{-1}$ exists complex numbers and proved the above fundamental property of complex numbers.

There Are More Irrationals than Rationals

The German mathematician **Georg Cantor** (1845–1918) used set theory to prove that there are more irrationals than rationals. You can see his idea already in children's books.

A *set* is just a collection A of objects, the objects in A, called its *elements*. For example, we have a set A of 7 apples and a set B of 7 bananas. For children, in some books the number 7 is represented by 7 apples $a_1, \ldots, a_7$ while in others by 7 bananas $b_1, \ldots, b_7$. Thus, as long as there is a one-to-one correspondence $a_i \leftrightarrow b_i$ between the elements of two sets A and B, they represent the same number, or have the same cardinality.

The sets of $\mathbb{Z}$ of integers $0, \pm 1, \pm 2$, A of even numbers $2n$ and B odd numbers $2n + 1$ with n in $\mathbb{Z}$, all have the same cardinality. The one-to-one correspondence between the elements of A and B for example, is given

$$A \to B \to A$$

$$2n \to 2n + 1 \to 2n.$$

A similar one-to-one correspondence proves that the cardinality $\mathrm{Card}(\mathbb{Z}) = \mathrm{Card}(\mathbb{N})$ where $\mathbb{N}$ is the set of natural numbers $1, 2, 3, \ldots$.

Given non-empty sets A and B, a *function* or *map* from A to B, written $f : A \to B$ is a rule f which assigns to a in A some $b = f(a)$ in B.

The set A is the *domain* and B is the *codomain* of f. Think of a function as a gadget with three components: domain, codomain, and the rule. For two functions $f_1 : A_1 \to B_1$ and $f_2 : A_2 \to B_2$ to be the same, we must have $A_1 = A_2$, $B_1 = B_2$, and for all a in $A_1 = A_2$, $f_1(a) = f_2(a)$.

A function $f : A \to B$ is *one-to-one* or *injective* if $f(a_1) = f(a_2)$ implies $a_1 = a_2$, i.e., no two a have the same value $f(a)$ in B. It is *onto* or *surjective* if each b in B is the *value* $f(a)$ of an a in A. In a familiar language, this means that B is the *range* of f, the set of all values $f(a)$ with a in A. Finally, f is *bijective* if it is both injective and surjective.

Let $\mathbb{R}$ be the set of real numbers. Consider the following four functions defined by the same rule $f(x) = x^2$, with $\mathbb{R}^\bullet$ denoting the set of non-negative reals; $f : \mathbb{R} \to \mathbb{R}$, $f : \mathbb{R} \to \mathbb{R}^\bullet$, $f : \mathbb{R}^\bullet \to \mathbb{R}$, and $f : \mathbb{R}^\bullet \to \mathbb{R}^\bullet$. The first one is neither injective ($f(-1) = f(1)$) nor surjective ($-1 = f(x)$ for no x in $\mathbb{R}$); the second one is surjective but not injective, the third one injective but not surjective. Only the fourth is bijective.

We denote by $\mathbb{Q}$ the set of rational numbers. The following is one of the most fundamental facts from set theory:

- $\mathrm{Card}(\mathbb{Q}) = \mathrm{Card}(\mathbb{N})$.

To prove this, it is enough to replace $\mathbb{Q}$ by $\mathbb{Q}^+$, the set of positive rationals.

Arrange the rationals $r_{ij} = \frac{a_i}{b_j}$ with GCD $(a_i, b_j) = 1$, $b_j = 1, 2, 3, \ldots$, and $a_i < a_k$ if $i < k$, as in Figure 5.3.

Now follow the arrows in Figure 5.3 to set up a bijective map f from the set $\mathbb{N}$ of natural numbers to that of positive rationals:

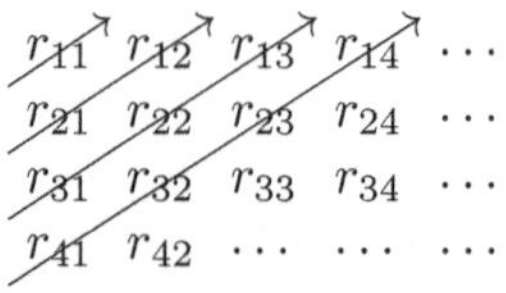

FIGURE 5.3: Arranging rationals in a sequence.

$f(1) = r_{11}$, $f(2) = r_{21}$, $f(3) = r_{12}$, $f(4) = r_{31}, \ldots$.

Definition. Given sets A and B, $\mathrm{Card}(B) > \mathrm{Card}(A)$ if there is an injection $f : A \to B$ but no surjection $g : A \to B$.

- $\mathrm{Card}(\mathbb{R}) > \mathrm{Card}(\mathbb{Q})$.

Proof. We may replace $\mathbb{R}$ by the open interval $(0, 1)$ because they have the same cardinality and $\mathbb{Q}$ by $\mathbb{N}$. Any real number x in the interval $(0,1)$ has a decimal representation $x = .d_1 d_2 d_3 \ldots$ It is unique if we choose one ending with all digits 0 and not 9 such as .5 and not .49999 . . ..

If $\mathrm{Card}(0, 1) = \mathrm{Card}(\mathbb{N})$, then we have a bijection f from $\mathbb{N}$ to $(0, 1)$. Suppose $f(n) = 0.d_{n1} d_{n2} d_{n3} \ldots$. Choose x in $(0, 1)$ with its jth digit different from d_{jj}, for every j. Then x is not $f(n)$ for any n. This contradicts the map is surjective. $\square$

Corollary. *There are more irrationals than rationals.*

A set S is *countable* if it is either finite or $\mathrm{Card}(S) = \mathrm{Card}(\mathbb{N})$.

Now that we have seen two infinities, $\mathrm{Card}(\mathbb{N})$ and $\mathrm{Card}(\mathbb{R})$, the later larger than the former, a natural question that arises immediately is: Is there an infinity strictly between these two? Cantor conjectured that there is none (*Continuum Hypothesis*) but could not prove it. In 1940, **Kurt Gödel** (1906–1978) proved that its truth or falsity cannot be proved within the axioms of set theory [Göd].

Part II

Geometry

6

Basic Geometry

The basic notions of length, area and volume were not alien to the prehistoric civilizations. The pyramids, palaces and great baths built more than 4000 years ago provide ample evidence for this. We begin our investigation of geometry with a discussion of area of simple geometric objects. What is meant by multiplying numbers, none of which is rational, is a delicate issue, as discussed in some detail in Chapter 5. Thus an area need not be a rational number. We derive the formula for the area of circle, *ab initio*, starting from scratch with the definition of the area of a rectangle.

Areas of Rectangles and Triangles

A *rectangle* is a plane figure formed by four straight lines meeting at right angles as shown below.

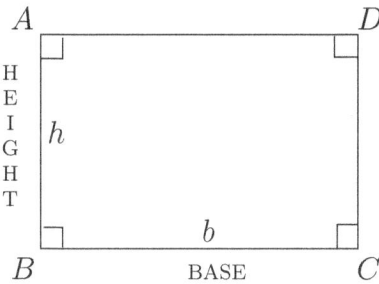

We define its area as:

$$\text{area of a rect } ABCD = \text{base} \times \text{height} = bh.$$

This is the most natural way to define area, for if we double the base then the area should double, or if we triple the height then the area should also triple. In modern language, we say that the area is linear in each variable b and h. Up to a constant $c > 0$, which depends on the units of measurements, this is the only definition of the area if we want it to have these obvious properties.

From this definition we can show that the area of a triangle is given by the formula

$$\text{area of triangle} = \frac{1}{2}(\text{base} \times \text{height}).$$

We consider three possible cases:

(1) The easiest case is that of a right triangle $\triangle ABC$. Obviously, its area is half the area of the rectangle $ABCD$,

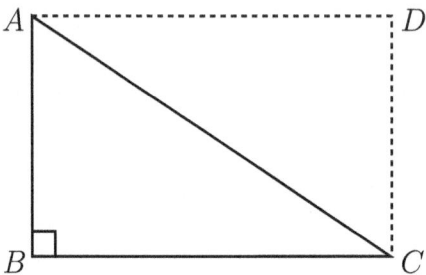

i.e.,

$$\text{area of } \triangle ABC = \frac{1}{2}(\text{area of the rectangle } ABCD)$$

$$= \frac{1}{2}(\text{base} \times \text{height}).$$

(2) To compute the area of an acute triangle $\triangle ABC$, we begin by dividing the triangle into two right triangles as shown below.

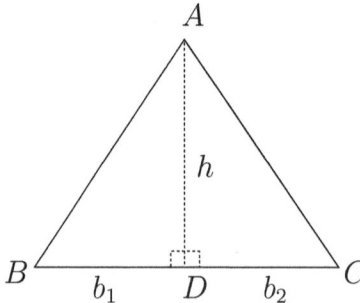

The base of $\triangle ABC$ is $b_1 + b_2$. Hence,

$$\text{area of } \triangle ABC = \text{area of } \triangle ABD + \text{area of } \triangle ADC$$

$$= \frac{1}{2}\, b_1 h + \frac{1}{2}\, b_2 h$$

$$= \frac{1}{2}\, h(b_1 + b_2)$$

$$= \frac{1}{2}\, \text{height} \times \text{base of triangle } ABC.$$

So, we have the same formula for the area of a triangle.

(3) Finally, the case of an obtuse triangle $\triangle ABC$.

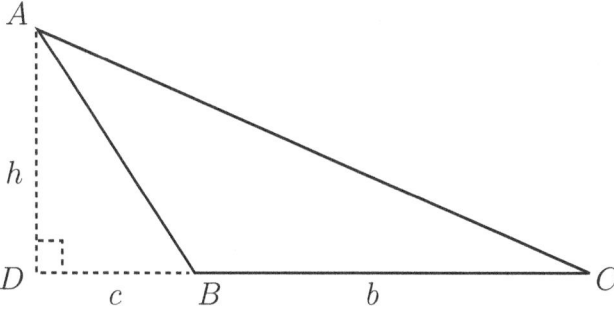

Now

$$\triangle ABC = \triangle ADC - \triangle ADB.$$

From this and the fact that $\triangle ADC$ and $\triangle ADB$ are right triangles,

$$\text{area of } \triangle ABC = \text{area of } \triangle ADC \ - \ \text{area of } \triangle ADB$$

$$= \frac{1}{2}\, h(b + c) - \frac{1}{2}\, hc$$

$$= \frac{1}{2}\, bh.$$

In the *Elements*, Book VI, Proposition 4, Euclid proves the following:

- *The sides of similar triangles are proportional.*

Proof. By placing the smaller triangle inside the bigger one, we may assume (see Chapter 8) that the triangles are ABC and ADE with $BC \| DE$, as shown in Figure 6.1.

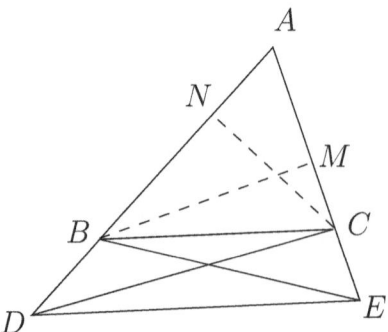

FIGURE 6.1: Proportionality of side lengths of similar triangles.

Let $|\triangle ABC|$ denote the area of a triangle ABC. Join B to E and C to D. Having the same base BC and being between the same parallels, the triangles so formed have equal areas, i.e.,

$$|\triangle BCD| = |\triangle BCE|. \tag{6.1}$$

Thus

$$\frac{|\triangle BCD|}{|\triangle ABC|} = \frac{|\triangle BCE|}{|\triangle ABC|}.$$

Suppose now that $BM \perp AE$ and $CN \perp AD$.

By using the formula for the area of a triangle, this gives

$$\frac{\frac{1}{2}\,BD\cdot CN}{\frac{1}{2}\,AB\cdot CN} = \frac{\frac{1}{2}\,CE\cdot BM}{\frac{1}{2}\,AC\cdot BM}$$

or

$$\frac{BD}{AB} = \frac{CE}{AC}.$$

By adding 1 to each side of this equation, we have

$$\frac{BD + AB}{AB} = \frac{CE + AC}{AC}$$

or

$$\frac{AD}{AB} = \frac{AE}{AC}.$$

Hence the proportionality of the corresponding sides of similar triangles ABC and ADE follows.   □

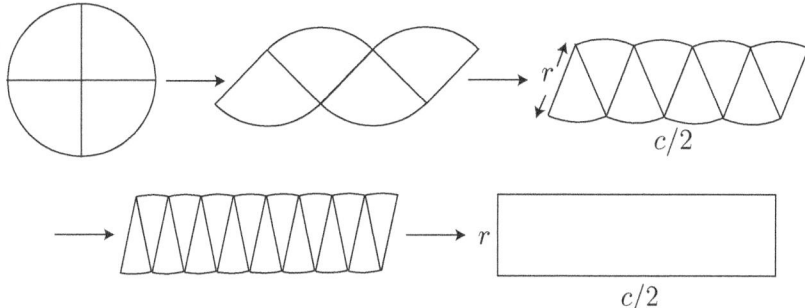

FIGURE 6.2: Changing a circle into a rectangle.

Area of Circles

Next we compute the area of a circle. The formula for the area of a circle was known to the Chinese and the Babylonians as half of the radius times the circumference. For this we cut a circle into a large and even number of equal slices and then place them as shown in Figure 6.2.

If n is very large, then this figure is almost a rectangle and exactly a rectangle as $n \to \infty$ with its base equal to half of the circumference and its height equal to the radius. So, using the formula for the area of a rectangle, we get the formula:

$$\text{area of circle} \; = \; \frac{1}{2} \; \text{circumference} \times \text{radius.} \tag{6.2}$$

It should be noted that one cannot claim the two sides of (6.2) to be exactly equal without having some idea of "limit." Thus these ancient civilizations were not unaware of the concept of limit.

To investigate this formula further, recall the last theorem in the previous section about the proportionality of the side lengths of similar triangles (triangles of the same shape). If we have two similar triangles with sides (x, y, z) and (X, Y, Z), as illustrated in Figure 6.3, then

$$\frac{X}{x} = \frac{Y}{y} = \frac{Z}{z} = \text{constant} = c, \text{ say}$$

or

$$X = cx$$
$$Y = cy$$
$$Z = cz,$$

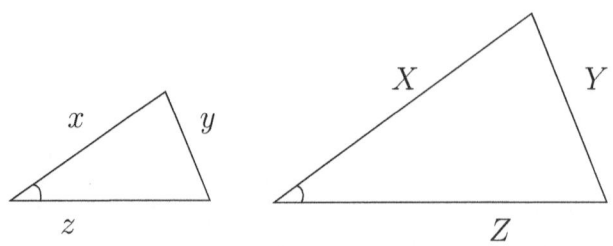

FIGURE 6.3: Similar triangles.

i.e., the sides of a triangle are proportional to the sides of any triangle similar to it.

Even though the proportionality of sides of similar triangles must have been obvious to the Egyptians from building the pyramids, the proof appeared much later in Euclid's *Elements*, Book VI, Proposition 4.

It is not hard to prove that all circles bear the same ratio of the circumference to the diameter, a fact well-known as far back as the Egyptians in 1650 BC.

- *The ratio circumference/diameter is the same for all circles.*

Proof. Let us consider any two (concentric) circles of different sizes. If C, R and c, r denote the circumference and the radius of the bigger and smaller circle, respectively, then it is enough to show that

$$\frac{C}{2R} = \frac{c}{2r}.$$

To approximate the circumferences, we inscribe regular n-gons inside the circles as shown in Figure 6.4 for $n = 8$.

We denote by C_n and c_n the perimeters of the polygons so formed, each of n equal sides. As illustrated in Figure 6.4, the parameters C_n and c_n approximate the circumference of each circle. But $c_n = n \cdot b_n$ and $C_n = n \cdot B_n$.

By the proportionality of sides of similar triangles, it follows that

$$\frac{c_n}{C_n} = \frac{nb_n}{nB_n} = \frac{b_n}{B_n} = \frac{r}{R}.$$

Thus,

$$\frac{c_n}{C_n} = \frac{r}{R}.$$

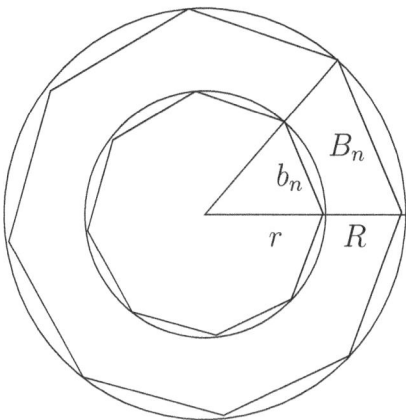

FIGURE 6.4: Definition of π.

Since $\frac{c_n}{C_n}$ is independent of n (it is always equal to the constant $\frac{r}{R}$) and since c_n approaches c and C_n approaches C as n increases, it follows that

$$\frac{c}{C} = \frac{r}{R}, \text{ or } \frac{c}{r} = \frac{C}{R}.$$

Therefore,

$$\frac{c}{2r} = \frac{C}{2R}. \qquad \square$$

Definition of π

Today we use the Greek letter π (pi) to denote this universal constant: any circle's circumference divided by its diameter. But the Greeks did not use this symbol. Although the first occurrence of the symbol π for this ratio appeared in 1706 in a book by William Jones, probably to initialize the Greek word περίμετρον (perimetron) for perimeter, its adoption in 1737 by Euler made it a standard symbol in mathematics.

Now we can easily deduce the well-known formula (famously proved by Archimedes) for the area of a circle with radius r, i.e.,

area of a circle of radius r is πr^2.

Proof. Since π is by definition the ratio circumference/diameter, we see that for a circle of radius r, its

$$\text{circumference} = \frac{\text{circumference}}{\text{diameter}} \cdot \text{diameter}$$

$$= \pi(2r)$$

$$= 2\pi r.$$

So by (6.2), the

$$\text{area} = \frac{1}{2} \text{ circumference} \times \text{radius}$$

$$= \frac{1}{2} \cdot 2\pi r \cdot r$$

$$= \pi \cdot r^2 \qquad \square$$

Pythagorean Theorem

This is a theorem central to all of geometry and multivariable calculus. There is the Fundamental Theorem of Algebra, the Fundamental Theorem of Arithmetic, and the Fundamental Theorem of Calculus. If there is a Fundamental Theorem of Geometry, it is the Pythagorean Theorem. Artifacts reveal that this theorem was known to the Babylonians and the Egyptians sometime around 2000 BC, and to the Chinese and Indians around 1500 BC. In fact, the oldest known proofs (or rather hints for proofs) of the Pythagorean Theorem go back to the Chinese and the Indians. However, the theorem is often referred to as the Pythagorean Theorem after the Greek mathematician Pythagoras (around 6th century BC) who neither discovered nor proved it. It is likely that his name was attached to it by his followers. The first complete proof by Euclid in Book I of his *Elements* is a masterpiece of how to write mathematics.

Among many known proofs, the one given by the Indian mathematician **Bhaskaracharya**, also spelled as **Bhaskara II** (AD 1114–1185), looks very easy, but assumes that the sum of three angles of a triangle is 180°, a fact which itself is equivalent to the Pythagorean theorem.

- (Pythagoras) *Given a right triangle with side-lengths a, b and hypotenuse c, we always have $a^2 + b^2 = c^2$.*

Proof. Bhaskar's proof is geometric, but it can be explained as follows: Cut a square of each side $a + b$ into four equal triangles and a square as shown in Figure 6.5. Since the sum of three angles of a triangle is 180°, the quadrilateral in the middle is actually a square.

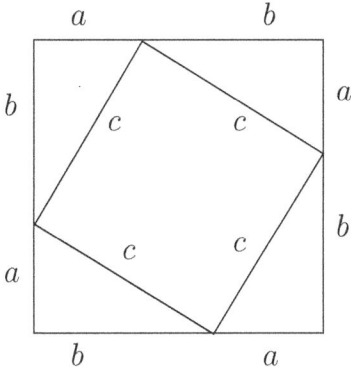

FIGURE 6.5: Bhaskar's proof of the Pythagorean theorem.

By using the formulas for the area of a rectangle and triangle, we see that $(a + b)^2 = 4(\frac{1}{2}ab) + c^2$, which simplifies to

$$a^2 + b^2 = c^2.$$ □

Third Proof ([Euc], Book VI, Prop. 31). In fact, proportionality of the side lengths of similar triangles implies the Pythagorean theorem. Let ABC be a right triangle with right angle at C and with sides a, b and c. From C draw perpendicular CD on AB, as shown in Figure 6.6.

Let $c = c_1 + c_2$ as shown. From three similar triangles, we have

$$\frac{a}{c} = \frac{c_2}{a} \text{ and } \frac{b}{c} = \frac{c_1}{b}$$

that is

$$a^2 = c \cdot c_2 \text{ and } b^2 = c \cdot c_1$$

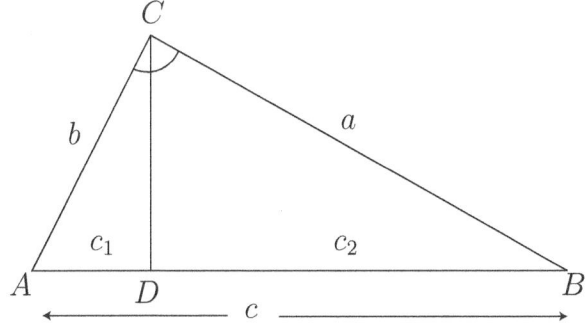

FIGURE 6.6: Euclid's second proof of the Pythagorean theorem.

which on adding gives

$$a^2 + b^2 = c(c_1 + c_2) = c^2. \qquad \square$$

Problems.

(1) If A, B, C with $0 < A \le B < C$ are the areas of equilateral triangles formed on the sides of a right triangle, show that $A + B = C$.

(2) Suppose a cord of length ℓ is tangent to the inner of the two concentric circles. Compute the area of the region bounded by the two circles (Figure 6.7).

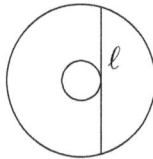

FIGURE 6.7: A problem on chord.

(3) As in Figure 6.8, circles of radii 6 and 2 are both tangent to line L and tangent externally to each other. Suppose the smaller circle rolls along the circumference of the larger one until it is again tangent to L, as shown. What distance has the center of the smaller circle traveled?

Pythagorean Triplets

A triplet (x, y, z) with x, y, and z all positive integers satisfying the equation $x^2 + y^2 = z^2$ is called a *Pythagorean triplet*. Note that $x^2 + y^2 = z^2$ if and only if $(nx)^2 + (ny)^2 = (nz)^2$ for any whole number n. We call the Pythagorean

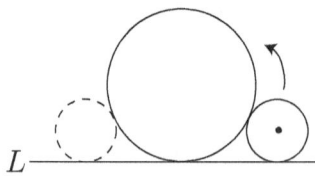

FIGURE 6.8: A problem on circumference.

triplet representing the smallest right triangle of a given shape a *primitive Pythagorean triplet*. One such triplet $(3, 4, 5)$ has been known since antiquity. Other commonly recognized triplets $(5, 12, 13)$ and $(8, 15, 17)$ were already known to Babylonians. The order of x, y is immaterial, but we shall write the odd one first. Both cannot be even if the triplet is primitive. The following algorithm describes them completely.

Theorem. *All the primitive Pythagorean triplets* (x, y, z), *i.e., integers* x, y, z *representing the side lengths of a right triangle (among similar ones of smallest size) are of the form*

$$x = a^2 - b^2$$
$$y = 2ab \qquad\qquad (6.3)$$
$$z = a^2 + b^2$$

where a, b *are integers such that*

(*i*) $a > b > 0$

(*ii*) a, b *are of opposite parity, i.e., one is odd, the other even*

(*iii*) a, b *have no common factor.*

Before we proceed with a proof, let's find some Pythagorean triplets (x, y, z) by using this algorithm.

Example. Let $a = 2$ and $b = 1$. Using equations (6.3) we get

$$x = 2^2 - 1^2 = 3$$
$$y = 2 \cdot 2 \cdot 1 = 4$$
$$z = 2^2 + 1^2 = 5,$$

Next we take $a = 3$ and $b = 2$. Our triplet (x, y, z) is now $(5, 12, 13)$ for

$$x = 3^2 - 2^2 = 5$$
$$y = 2 \cdot 3 \cdot 2 = 12$$
$$z = 3^2 + 2^2 = 13,$$

By choosing two numbers a and b as above, we can generate up to proportionality all right triangles with sides whose lengths are whole numbers. Thus there are infinitely many primitive Pythagorean triplets. Now that we have seen how powerful the theorem is, we sketch a proof.

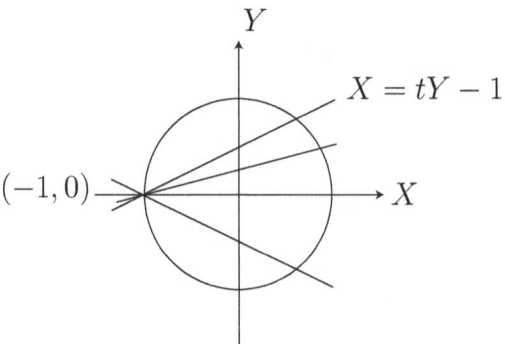

FIGURE 6.9: Parameterizing the circle.

Proof of the Theorem. To begin, we write the equation $x^2 + y^2 = z^2$ in the form

$$\left(\frac{x}{z}\right)^2 + \left(\frac{y}{z}\right)^2 = 1, \tag{6.4}$$

so that we can see that finding the Pythagorean triplet is equivalent to finding points (X, Y) on the circle

$$X^2 + Y^2 = 1 \tag{6.5}$$

with coordinates $X = \frac{x}{z}$ and $Y = \frac{y}{z}$ as rational numbers in the lowest form. (Recall that a *rational number* is a ratio of two whole numbers such as $\frac{1}{2}$, $\frac{3}{5}$, $\frac{22}{7}$.)

From analytic geometry, we know that the circle given by (6.5) can be parameterized as (see Figure 6.9)

$$\begin{aligned} X &= \frac{t^2 - 1}{t^2 + 1} \\ Y &= \frac{2t}{t^2 + 1}. \end{aligned} \tag{6.6}$$

To get points with rational coordinates, we plug $t = \frac{a}{b}$ where a, b are integers in the lowest form into (6.6) and solve for X and Y:

$$X = \frac{\left(\frac{a}{b}\right)^2 - 1}{\left(\frac{a}{b}\right)^2 + 1}$$

which simplifies to

$$\frac{a^2 - b^2}{a^2 + b^2}$$

and

$$Y = \frac{2\left(\frac{a}{b}\right)}{\left(\frac{a}{b}\right)^2 + 1}$$

FIGURE 6.10: Egyptian rope protractor.

which simplifies to

$$\frac{2ab}{a^2 + b^2}.$$

Recall that $X = \frac{x}{z}$ and $Y = \frac{y}{z}$. Since $t = \frac{a}{b}$ is in the lowest form, so are X and Y above. Hence

$$x = a^2 - b^2,$$
$$y = 2ab,$$
$$z = a^2 + b^2.$$

It can now be checked that a and b must be as stated. □

Recall that all the Pythagorean triplets are obtained from the primitive ones (x, y, z) as (nx, ny, nz) for n in $\mathbb{N}$.

Problems.

1. Derive the parameterization (6.6) for the circle (6.5).

Hint: Intersect (6.5) with the family of lines $X = tY - 1$ through the fixed point $(-1, 0)$ on (6.5). In other words, substitute $X = tY - 1$ in (6.5) and solve for Y and X in terms of t, the slope of the line $X = tY - 1$ of this family, as shown in Figure 6.9.

2. Find a, b which on substitution in (6.3) gives $x = 12,709, y = 13,500$ and $z = 18,541$.

Remarks.

1. The sophistication of the Babylonian mathematics is remarkable. (See [Hør].) During the period 1900–1600 BC, the Babylonians (see Figure 6.11) compiled tables of Pythagorean triplets (see [Wei-1, pp. 8–10] and also Figure 6.12.) It is hard to believe that they stumbled upon the primitive Pythagorean triplets like $x = 12,709$, $y = 13,500$, and $z = 18,541$. Plausibly, they were aware of the algorithm provided by the theorem above for generating the primitive Pythagorean triplets.

FIGURE 6.11: Map of the region where the Babylonian civilization flourished (modern-day Iraq).

2. It is believed that as far back as 5000 years ago the Egyptian rope-stretchers used the Pythagorean triplet $(3, 4, 5)$ to construct right angles such

FIGURE 6.12: Plimpton 322 is the Babylonian mathematical tablet numbered 322 in the collection of G. A. Plimpton housed in Columbia University. Dated around 1800 BC, it was recovered from the desert in Iraq.

as for pyramid-building. In order to construct a triplet $(3, 4, 5)$, a rope was divided by eleven knots into twelve equal segments. Since the size of the rope is arbitrary, the Egyptians must have known the proportionality of similar triangles. The rope was laid and stretched on the ground to make a right triangle with side lengths 3, 4, and 5 as shown in Figure 6.10. Taking the rope long enough eliminated any error in measurement.

7

Greece: Beginning of Theoretical Mathematics

The pure or deductive mathematics, as we know it today, is a gift from Greece. The classical Greek period is roughly the millennium 600 BC–AD 400. During this period, the Greeks were spread over a territory that included present-day Greece, some colonies in south Italy such as the important city of Syracuse in Sicily, western Turkey, as well as northern Egypt, in particular, the city of Alexandria. This great city was founded by Alexander the Great in 322 BC. It rivaled, and at times even surpassed, Athens as the intellectual center of Greek civilization.

The Greeks built their mathematics on what they had learned from the earlier civilizations of Egypt and Babylonia. These territories were either incorporated into Greece or were neighboring lands the Greeks would often visit. Although the Greeks learned a substantial amount of mathematics from Egypt and Mesopotamia, theoretical mathematics originated with the Greeks. The creation of this mathematics, which transcends practical needs, was one of the most remarkable events in the history of mankind. It had an immense impact on the development of all sciences.

Thales (624–547 BC) from Miletus was the first recognizable Greek mathematician. He is believed to have traveled extensively and learned geometry from Egyptians and astronomy from Babylonians. He predicted the solar eclipse of 585 BC. It is believed that Thales was the first mathematician to insist on logical proofs based on deductive reasoning rather than experiment and intuition to support an argument. It is for this reason that some consider him as the father of pure mathematics. In geometry his name is connected with propositions such as:

(i) the base angles of an isosceles triangle are equal;

(ii) if the side and the two adjacent angles (angle-side-angle) of one triangle are equal to that of another, the two triangles are *congruent*, that is, are identical in size and shape;

(iii) any triangle inscribed in a semicircle is a right triangle;

(iv) the sides of similar triangles are proportional.

To see exactly what Thales might have proved, see the commentary that follows these theorems in [Euc].

A better-known figure is **Pythagoras** (572–497 BC). It is believed that he spent a considerable part of his life in Egypt and Babylonia. It is quite likely that he was a student of Thales. Pythagoras had many disciples known as Pythagoreans who formed a semi-religious school. They called their discipline "mathema," that is, the art of learning. The roots of our own word "mathematics" go back to it. They divided their discipline into four branches: arithmetic, astronomy, geometry and music. The **quadrivium**, that is, these four subjects, formed the core of medieval and later Western higher education. Their research delved, among other things, into polygonal numbers (e.g., triangular numbers, pentagonal numbers, etc.). It is generally believed that Pythagoreans knew about the irrationality of $\sqrt{2}$ but did not accept it. However, the irrationality of the golden ratio $(1 + \sqrt{5})/2$ seems to be older than that of $\sqrt{2}$.

Pythagoras was followed by **Plato** (427–347 BC), who established the Academy in Athens. It is said that he had a sign on its gate saying, "Let no one ignorant of geometry enter here." After the demise of the Pythagorean school, its followers moved to the Academy where they continued to promote mathematics. Another prominent school at that time was Eleatic whose most famous member was **Zeno** (490–430 BC), known for several paradoxes in logic.

One of the most famous names in Greek mathematics, in fact in all of mathematics, is **Euclid** (around 300 BC). He compiled the mathematics that was known in his time into thirteen books called the *Elements*. (See Figure 7.1.) Without a doubt, these books are the most important mathematical texts of all time. In fact, no other book seems to have such a universal appeal as Euclid's *Elements*. In India and China, the two countries that account for almost half of humanity, Euclid is taught in schools.

The **Mouseion** or Museum is the name given to the library and university at Alexandria. This complex was founded around 300 BC, by which time the center of Greek culture had already moved to Alexandria. The library at the Mouseion is said to have possessed over half a million volumes. It is here at the Mouseion that Euclid compiled his *Elements*, which were to become the model for millennia for mathematical writings. Even N. Bourbaki's *Elements of Mathematics*, which have greatly influenced the mathematics of our own time, were modeled after Euclid's *Elements*. Euclid was the first mathematician, or at least the first one whose works have been reconstructed, to formalize the mathematical knowledge of his time in such a way that even today it is still the way to write mathematics.

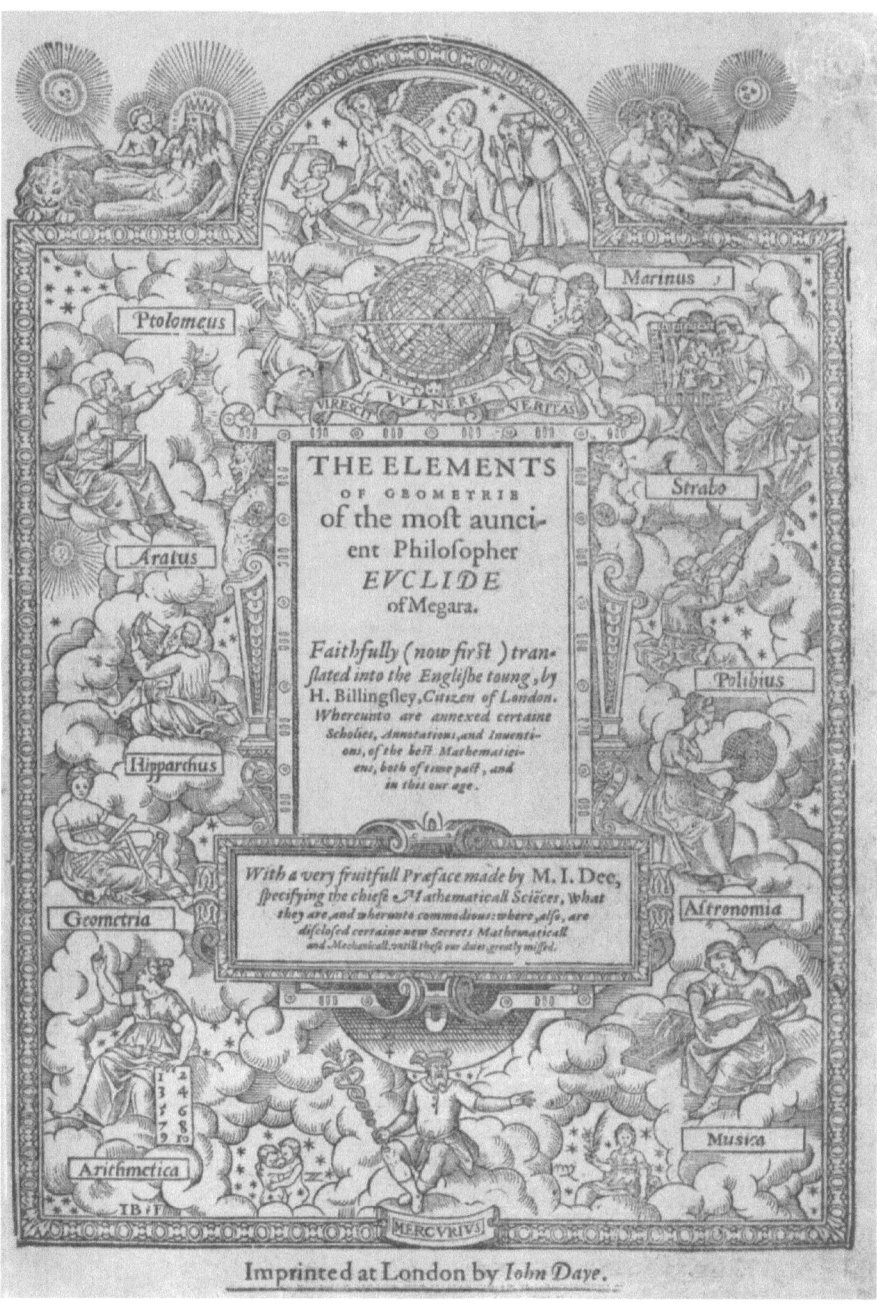

FIGURE 7.1: Title page of Sir Henry Billingsley's first English version of Euclid's *Elements* published in 1570.

The Golden Age of Greek Mathematics

The third century BC and soon thereafter is considered to be the golden age of Greek mathematics. The works of some great mathematicians from this period are still taught as mainstream mathematics. For example, the chapters on conic sections in analytic geometry books are a reformulated version (in Cartesian coordinates) of the *Books on Conics* by **Apollonius** (250–175 BC), (see [Too]). It is Apollonius who is often credited with inventing the terms parabola, ellipse and hyperbola, meaning exact, deficient and excessive in that order. **Claudius Ptolemy** (AD 100–178) compiled the result of centuries of knowledge from Babylonian astronomers and Greek geometers into a book on astronomy. This book, titled *Almagest* (a Latin distortion of its Arabic name, which means "the greatest") was to become a bible on astronomy for centuries to come.

Archimedes of Syracuse (287–212 BC) was the greatest mathematician, scientist and engineer of antiquity, and one of the greatest of all times. Isaac Newton, not known for modesty, had this to say of Archimedes (and Galileo Galilei): "If I have seen farther than others, it is because I stood on the shoulders of giants." Archimedes was born in Syracuse, where he spent most of his life and where he died. He traveled to Egypt to study at the Mouseion in Alexandria. During his stay in the Nile Valley, he invented the so-called *Archimedes screw*. It is a long cylinder with a tightly fitting screw. When the screw is rotated inside the cylinder it pulls water through the cylinder. Archimedes was so absent-minded that he would forget to eat and bathe. The most famous story about Archimedes concerns the crown of King Hieron of Syracuse, who was suspicious of the goldsmith commissioned to make the crown. Archimedes was asked to determine without breaking the crown if it was of pure gold or just gold-plated. Archimedes wrestled with the problem for a long time until one day during one of his rare baths he suddenly hit upon the solution. Jumping from the bath, he ran through the street shouting, "Eureka, Eureka!" Unfortunately, he was so absorbed in his thoughts that he forgot to put on his clothes. The solution of the crown problem culminated in his Law of Floating Bodies.

Law of Floating Bodies. *When an object (or body) floats in a liquid, the weight of the body is equal to the weight of the liquid it displaces.*

Equally well known is his Law of Levers (see figure below).

Law of Levers. *If a (weightless) bar is balanced on a fulcrum with weights at each end as shown in the figure, then* $wL = lW$.

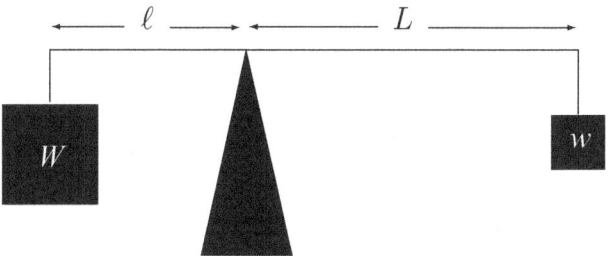

He formulated a similar law for pulleys.

When Romans under Marcellus attacked Syracuse, Archimedes was called upon to help defend his city. Putting his discoveries about levers and pulleys to work, he bombarded the enemy with rocks so huge that the bewildered enemy fled in terror. Finally, however, through treachery, the Romans were able to enter the city. Even though Marcellus gave explicit orders to spare Archimedes, a soldier slew Archimedes while he was deeply absorbed in solving a mathematical problem.

Archimedes contributed to almost every branch of science. In pure mathematics, for example, he computed the area of a circle, the area of a parabolic segment, and the surface area and volume of a sphere (all without the Fundamental Theorem of Calculus, an astonishing feat); he studied space curves; he developed the formula generally credited to Heron which expressed the area of a triangle in terms of its three side lengths (see below); his cattle problem leads to a Pell equation whose solution could be found only two millennia later with a computer search; and he determined the fractions with smallest denominators which most closely approximate the number π (the most well known $\pi \approx \frac{22}{7}$).

A contemporary of Archimedes is **Eratosthenes** of Alexandria (275–195 BC). He knew the earth was round and computed its radius. He is best known for the *sieve of Eratosthenes*. This is an algorithm to find all prime numbers up to a given bound N. We list all the integers, starting with the first prime 2, up to N, say $N = 20$.

$$2 \ 3 \ \cancel{4} \ 5 \ \cancel{6} \ 7 \ \cancel{8} \ \cancel{9} \ \cancel{10}$$
$$11 \ \cancel{12} \ 13 \ \cancel{14} \ \cancel{15} \ \cancel{16} \ 17 \ \cancel{18} \ 19 \ \cancel{20}$$

First delete all non-trivial multiples of 2, namely $4, 6, 8, \ldots$. The first number next to 2 to survive deletion, that is 3, is the next prime. Now delete all non-trivial multiples of prime 3 from the remaining list. We see that 5 is the next prime. And this is all we have to do, for if a number $\leq N$ is not a prime it has a prime factor $p \leq$ the integer part of $\sqrt{N}$, e.g., $p \leq 4$ for $N = 20$. So our list of primes up to 20 is

$$2, 3, 5, 7, 11, 13, 17, 19.$$

At the beginning of the first millennium AD, **Heron** (1st century AD) had the *Heronic formula* for the area of a triangle with side lengths a, b and c. The quantity

$$s = \frac{a+b+c}{2}$$

is the *semi-perimeter* of the triangle. The area A of this triangle is given by

$$A = \sqrt{s(s-a)(s-b)(s-c)}. \tag{7.1}$$

Note that, from the triangle inequality, $s \geq a, b, c$. Hence all the factors under the square root are ≥ 0. It is worth pointing out here that this is a special case of a formula of **Brahmagupta** (c. AD 598-668) for the area of a quadrilateral of sides a, b, c and d inscribed in a circle. Again

$$s = \frac{a+b+c+d}{2}$$

is the *semi-perimeter* of the quadrilateral. Brahmagupta's formula for the area A of this quadrilateral is

$$A = \sqrt{(s-a)(s-b)(s-c)(s-d)}. \tag{7.2}$$

We remark that:

(i) It is unreasonable to expect the factor s also in (7.2). For (7.2) to represent an area, it has to be a two-dimensional object. The square root of four linear factors is precisely a two-dimensional object. Likewise, the square root of a non-negative product of six linear factors would represent a volume.

(ii) If in formula (7.2) we take $d = 0$, the quadrilateral becomes a triangle with sides a, b and c, and the formula (7.2) reduces to formula (7.1). The formula (7.1) appears in the book *Metrica* of Heron. Heron sometimes gave proofs, but always his aim was to calculate. As suggested above, the formula (7.1) itself is due to Archimedes.

Diophantus and Number Theory

Every algebraic geometer and number theorist is familiar with the name **Diophantus** of Alexandria (AD mid-3rd century). His work titled *Arithmetica,* of which six of the original thirteen books remain, influenced Pierre de Fermat generally regarded as the father of modern number theory. In fact, there are important branches of number theory named after Diophantus, called Diophantine equations, Diophantine geometry and Diophantine approximations. Diophantine equations are polynomial equations in two or more variables with integer coefficients. The solutions are also sought in integers. We give a few examples of some famous *Diophantine equations.*

(1) **Aryabhat-Brahmagupta equation.** This is commonly called the Pell equation, but it has essentially nothing to do with Pell. It was Euler who first referred to it as Pell's equation, the subject of his correspondence with an Englishman named John Pell. This Diophantine equation can be traced back to Archimedes. His complicated "*cattle problem*" which he had sent to Eratosthenes (see [Cal, p. 151]) leads to the equation

$$x^2 - 4,729,494y^2 = 1.$$

A non-trivial solution $(y > 0)$ to this equation was found only in 1974 by a computer search. The existence of such a solution is guaranteed by the Dirichlet Unit Theorem in algebraic number theory (see [Cha-1], p. 78).

In general, let $m = 2, 3, 5, 6, \ldots$ be a positive integer with no square factor > 1. The equation
$$x^2 - my^2 = 1,$$

called *Pell's equation*, has been studied since antiquity. We shall return to it in a later chapter.

(2) **Pythagorean triplets.** We are already familiar with the Diophantine equation
$$x^2 + y^2 = z^2$$

in connection with the Pythagorean triplets such as $(3, 4, 5)$, $(5, 12, 13)$, $(8, 15, 17), \ldots$. Another way to rephrase the above equation is: what squares are the sum of two squares?

(3) **Fermat's Equation.** As has already been said, Diophantus' *Arithmetica* influenced Fermat who in 1637 asked: what cubes are the sum of two cubes? More generally, he studied the Diophantine equation, called *Fermat's equation*

$$x^n + y^n = z^n \quad (n \geq 3). \tag{7.3}$$

Fermat claimed in the famous margin of his copy of *Arithmetica* of Diophantus (see Figure 7.2) that he had discovered a marvelous proof that the equation (7.3) has no solution in integers x, y and z (with $xyz \neq 0$) "but this margin is too small for it." It is anybody's guess if Fermat really had a proof. Nevertheless, this claim has always been known as **Fermat's Last Theorem**. Actually, according to Dickson [Dic], in 972, **al Khujandi** (940–1000) (from Khujand in Tajikistan) had already considered equation (7.3) for $n = 3$ and gave a defective proof of Fermat's Last Theorem for this exponent. Various attempts to "reprove" this theorem resulted in great advances in number theory, including the birth of new branches of number theory such as algebraic number theory.

Fermat's Last Theorem was proved in 1993 by Sir Andrew Wiles, building on the previous work by several mathematicians, e.g., G. Frey, J.-P. Serre, and K. Ribet to mention a few.

Remark. There are other kinds of equations studied in number theory, where the variables appear as exponents. The variables and the constants are supposed to represent whole numbers. One such equation is

$$a^x - b^y = 1.$$

The *Catalan conjecture* says that this equation with a, b, x, y all > 1, is solvable in integers only when $a = 3, b = 2$, in which case it has only one solution, $x = 2, y = 3$. It was proved by Mihailescu in 2002.

Finally we mention **Pappus** (around AD 300). From what we know about Greek history, he was the last creative Greek mathematician. His eight books of *Synagoge*, better known as the *Mathematical Collection*, still exist. For example, Book 7 describes "Analysis," a kind of deductive reasoning. Although the *Synagoge* is not of the same quality as the earlier Greek classics of the Alexandrian school, it is an indispensable record of parts of mathematics that would have otherwise been lost. These books had a great influence on Descartes, whose invention of the Cartesian coordinates supplied the missing link between algebra and geometry. For details, see [Jon].

Demise of Greek Mathematics

During the first three or four centuries AD, as it often happens, the Christians went from oppressed to oppressors themselves. The Greeks who were not Christian were branded pagan, or evil. Christianity became the state religion and the worship of Greek gods was banned. Christian zealots killed those who refused to convert. Such was the fate of **Hypatia** (355–415), an eminent mathematician in Alexandria. Her father **Theon** was himself a mathematician of great repute. She wrote a brilliant commentary on *Arithmetica* of Diophantus. Hypatia's death is generally regarded as the end of Greek mathematical tradition in Alexandria and elsewhere in Greece. The remaining mathematicians fled to Constantinople, the capital of the Eastern Empire and later from there to Arabia. Their impact on the Arabic mathematics will be discussed in Chapter 14. The Mouseion had already been defiled in 392 and finally burned down in 641. The Academy in Athens was closed in 529 by order of Emperor Justinian.

DIOPHANTI
ALEXANDRINI
ARITHMETICORVM
LIBRI SEX,
ET DE NVMERIS MVLTANGVLIS
LIBER VNVS.

CVM COMMENTARIIS C. G. BACHETI V. C.
& obseruationibus D. P. de FERMAT Senatoris Tolosani.

Accessit Doctrinæ Analyticæ inuentum nouum, collectum
ex varijs eiusdem D. de FERMAT Epistolis.

TOLOSÆ,
Excudebat BERNARDVS BOSC, è Regione Collegij Societatis Iesu.

M. DC LXX.

FIGURE 7.2: Arithmetica of Diophantus.

8

Euclid: The Founder of Pure Mathematics

In the previous chapter, we discussed the contributions of various Greek mathematicians to mathematics without examining their mathematics. This chapter is devoted to studying the Greek mathematics, especially of Euclid. Recall that Euclid compiled the mathematics known at his time in a style that is still considered to be the benchmark for rigor and precision. It will not be inappropriate to call Euclid the founding father of theoretical mathematics. According to Euclid, a meticulously written work of mathematics has the following essential components:

(i) Carefully chosen definitions to name and make precise the objects of one's study.

(ii) Self-evident facts, called postulates or axioms, that cannot be proved, so must be taken for granted.

(iii) Theorems or facts that are derived only from the postulates and previously proven facts.

We illustrate these components by a brief discussion of Euclid's Book I, which deals with plane geometry. The book begins with simple definitions and postulates and culminates in the Pythagorean theorem, which may be considered as the Fundamental Theorem of Geometry. For details, see [Euc].

Euclid's Proof of Pythagorean Theorem

We now sketch Euclid's proof of this fundamental theorem, a masterpiece in the art of proof writing. We skip altogether what Euclid calls common notions, such as doubles of equals are equal. Labels for propositions, such as Proposition I.47, refer to the 47th proposition in Book I. We take up only the most prominent definitions.

Definitions

Euclid chooses definitions that are easy to use and least controversial. For example, a right angle can be defined in one of the following two ways below.

Definition A When a straight line falls on another one in such a way that the two angles are equal, then each is called a *right angle* as shown in Figure 8.1.

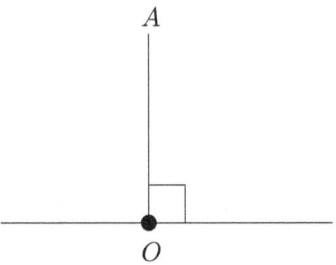

FIGURE 8.1: Correct definition of a right angle.

Definition B When two straight lines cross in such a way that all the four angles are equal, then each is called a *right angle* as shown in Figure 8.2.

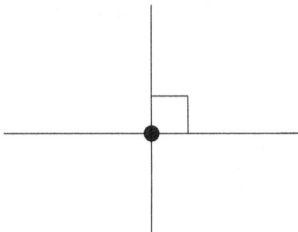

FIGURE 8.2: Wrong definition of a right angle.

Of course, Euclid chooses Definition A (listed as Def. 10 in Book I). In Definition B there is no guarantee the two lines will ever manage to cross in such a way that the four angles are equal. However, in Definition A all one has to do is to move line OA, keeping the end O fixed, clockwise or counterclockwise until the two angles are equal. (Note the obscure assumption on continuity to guarantee that, at some stage, the two angles will be equal.) Now we list some of Euclid's other definitions. For the rest, see [Euc, vol. I, pp. 53–54]. The numbering is as in Euclid's Book I. If necessary for clarity, we shall feel free to use the current terminology as follows:

1. A *point* is a part of the plane that cannot be further subdivided.

15. A *circle* is the set of points in the plane equidistant from a fixed point. This fixed point is the *center* and the equidistance is the *radius* of this circle.

24. An *equilateral* triangle is a triangle with all sides equal.

25. An *isosceles* triangle is a triangle with two equal sides.

27. A *right triangle* is a triangle with a right angle.

35. In the plane, two straight lines are *parallel* if they never meet.

Remarks on Terminology and Notation

1. Note that Euclid, in Book I does not work with areas, which are numbers, to prove the theorem.

When Euclid says two figures are equal, he means they are equal in the sense that each can be cut and placed on the other in such a way as to cover it completely.

2. The word congruent is used primarily for triangles. Two triangles are *congruent* if they are not only equal, but also similar in shape. Flipping is allowed. For example, the following two triangles, each of sides 2, 3 and 4 units of length, are congruent.

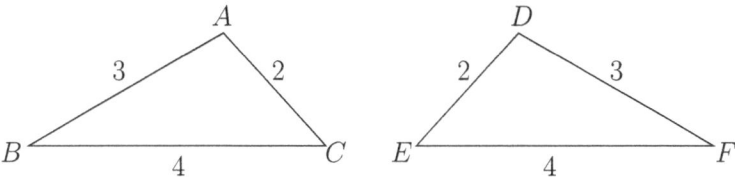

3. If line AB is parallel to line CD, we shall write $AB\|CD$.

4. Similarly for AB perpendicular to CD, we write $AB \perp CD$.

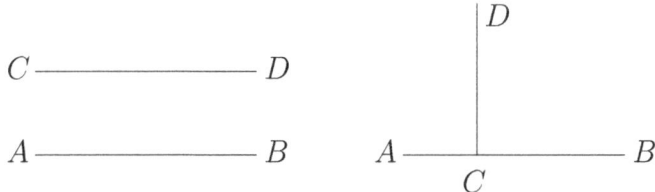

Postulates

The *postulates* are self-evident facts that cannot be proved, hence they must be taken for granted. Here are all the postulates from Book I. The numbering is again traditional. We rephrase some of them so as to reflect what Euclid really meant in the current language of mathematics.

1.   *One and only one straight line can be drawn between two points.*

2.   *A terminated straight line can be extended on each end to any length.*

3.   *One and only one circle can be drawn with a given radius and center.*

4.   *All right angles are equal to each other.*

5.   **The Fifth Postulate.** *If a straight line meets two other straight lines in such a way that the sum of two interior angles formed on one side is less than two right angles, then these two lines meet on that side (see Figure 8.3).*

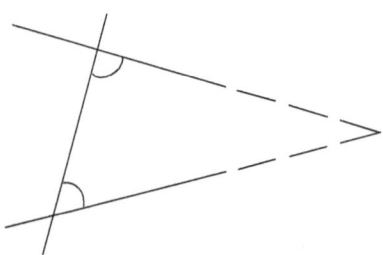

FIGURE 8.3: Fifth postulate.

Propositions

There are 48 propositions in Book I. Proposition 47 is the famous Pythagorean theorem, the next one is its converse. To prove Proposition I.1, we are allowed to use only the postulates, but for any subsequent proposition, we may use the propositions that have already been proved. We state only the basic ones, rarely supplying proofs. For a complete account see, [Euc]. We mark the important propositions as theorems. The abbreviation Q.E.D. stands for the end of proof (*quod erat demonstrandum*, the Latin phrase meaning "which was to be demonstrated").

Proposition I.1. *It is always possible to draw an equilateral triangle on a given finite straight line.*

Proof. Let AB be the given line. Draw two circles (Postulate 3) with centers at A and at B, but of the same radius AB. If C is a point of their intersection, join C to A and to B. (This is possible by Postulate 1.) Then ABC is an equilateral triangle (see Figure 8.4).   □

Remark. Letting mathematical considerations override historical ones, we let modern mathematical thinking creep into our presentation. Euclid would only

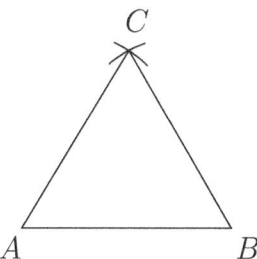

FIGURE 8.4: Constructing equilateral triangles.

say "to draw a straight line between two points" or "to construct an equilateral triangle." Euclid's approach is rather constructive, not existential. There are some minor flaws in some of Euclid's proofs. For example, how do we know that the two circles intersect at C and D? These issues were addressed by **David Hilbert** (1862–1943) in [Hil-1]. See also [Har].

The next proposition is usually referred to as SAS (side, angle, side).

Proposition I.4. *If any two sides of one triangle are equal to the respective two sides of another triangle and so are the angles between them, then these two triangles are congruent.*

Proof. Let ABC and DEF be two triangles with sides $AB = DE$, $AC = DF$ and the angle $\angle A = \angle D$, as in Figure 8.5.

Being equal, we can place AB on DE so that A falls over D and B over E. Since $\angle A = \angle D$, by definition of equality of angles, the line AC runs along DF. Now since $AC = DF$, C falls over F. By Postulate 1, BC has to coincide with EF. □

Remark. The issue of superposition in the proof of Proposition 4 needs to be addressed. How do we know that we can place line AB on DE? There is no

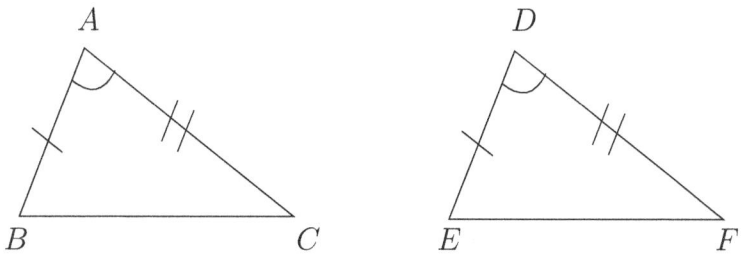

FIGURE 8.5: SAS Proposition.

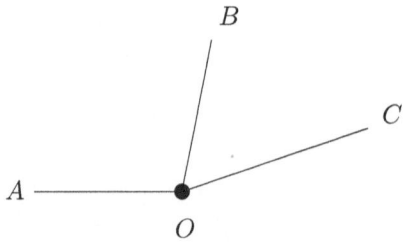

FIGURE 8.6: Proposition I.13–14.

postulate that says we are allowed to do this. To appreciate the genius that Euclid was, the reader is challenged to prove that as a consequence of the first three postulates, we can superimpose line AB on line DE. Euclid shows how to do this in Propositions 2 and 3.

Proposition I.5. *The two angles of an isosceles triangle at the base are equal.*

Proof. Exercise (or see [Euc, vol. I, p. 286]). ◻

Proposition I. 13–14. *Let three straight lines AO, BO and CO, meet at O as shown in Figure 8.6. Then AOC is a straight line if and only if the sum $\angle AOB + \angle BOC =$ two right angles.*

Theorem I.20. (Triangle inequality) *Any two sides of a triangle are together greater than the third one.*

Proof. Exercise or see [Euc, vol. I, p. 286]. ◻

Theorem I.29. *Suppose a straight line crosses two parallel lines as shown in Figure 8.7. Then the alternating angles are equal, and so are the corresponding angles. That is $\angle 2 = \angle 3$ and $\angle 1 = \angle 3$, respectively.*

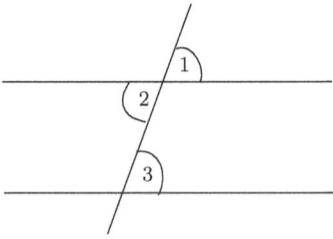

FIGURE 8.7: Proposition I.29.

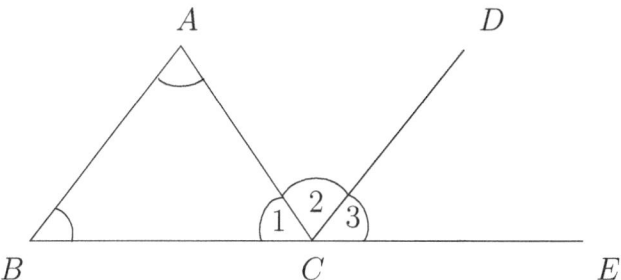

FIGURE 8.8: Sums of angles of a triangle.

Proposition I.31. *It is always possible to draw a straight line through a given point and parallel to a given line.*

Theorem I.32. *The sum of three angles of a triangle is always two right angles.*

Proof. Let ABC be the given triangle. By Postulate 2, extend BC to E and by Proposition I.31, draw $CD\|BA$ (Figure 8.8). By Theorem I.29, $\angle A = \angle 2$, $\angle B = \angle 3$. And of course, $\angle C$ is the same as $\angle 1$. Hence $\angle A + \angle B + \angle C = \angle 1 + \angle 2 + \angle 3 =$ two right angles (by a slight generalization of Proposition I.13–14). $\qquad\Box$

Proposition I.41. *If a parallelogram and a triangle are on the same base and between the same parallels, the parallelogram is twice that of the triangle.*

Theorem I.47. (Pythagoras) *For a right triangle, the square on the hypotenuse is equal to the sum of those on the other two sides.*

Proof. (Euclid). Let ABC be the right-angled triangle with BC as the hypotenuse, as shown in Figure 8.9.

Join A to D and G to C (Postulate 1). Draw from A line $AE\|BD$ to form rectangle $BDEF$ (Proposition I.31). We shall show that the rectangle $BDEF =$ the square $ABGH$.

First, by Proposition I.4 (SAS),

$$\Delta ABD = \Delta BCG. \tag{8.1}$$

By Proposition I.13–14, HAC is a straight line. Hence ΔBCG and square $ABGH$ are on the same base and between the same parallels. Therefore, by Proposition I.41, the square $ABGH$ is twice the triangle ΔBCG. Similarly

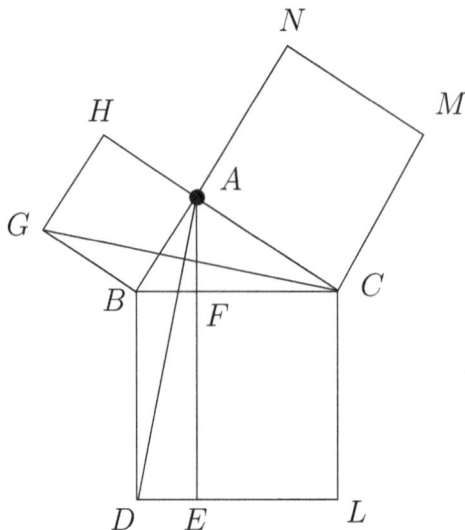

FIGURE 8.9: Proof of the Pythagorean theorem.

the rectangle $BDEF$ is twice the triangle $\triangle ABD$. Consequently by (8.1), the square $ABGH =$ the rectangle $BDEF$. It can be proved in a similar way that the square $ACMN =$ the rectangle $ELCF$. Since these two rectangles add up to the square on the hypotenuse, the theorem follows.   □

Some Comments on Euclid's Proof

There are numerous proofs of the Pythagorean theorem, all interesting in their own right. It may be stressed that Euclid's proof of I.47 is considered to be original, because he wanted to include this theorem in his book as early as possible. Therefore, he did not use earlier known proofs that used similarity. Euclid does not discuss proportionality until Book V where the ratio is defined for the first time (Definition 3, Book V). Only in Proposition VI.4 does he prove the proportionality of the sides of similar triangles, which leads to another proof of this theorem.

The proof of the Pythagorean theorem by Bhaskar in Chapter 6 seems to be a favorite among students. A proof, or at least a hint for a proof, of this theorem was known to the ancient Chinese. The third proof we gave in Chapter 6 may be the original one by Pythagoras [Euc, vol. I, p. 353] and is based on the proportionality of the side lengths of similar triangles. However, as we discussed in Chapter 5, the concept of proportionality is a subtle one. The ratio a/m when m is a positive integer has an obvious meaning: divide a into m equal parts. But what do we mean by $\sqrt{2}/\pi$? It is worth mentioning that **Eudoxus** (408–355 BC) developed a general theory of proportion to explain the ratio of quantities that are not rational, such as $\sqrt{2}/\pi$. Eudoxus' theory, which can be found in Book V of Euclid, is astonishingly modern in dealing with such issues. Even Richard Dedekind, whose construction of real numbers using cuts is now a standard definition of reals, acknowledges that the work of Eudoxus has been a source of his theory (of Dedekind cuts). Eudoxus is also credited with turning astronomy into a mathematical science.

Non-Euclidean Geometry

Euclid believed that his fifth postulate was unnecessary. In other words, he was convinced that it should be a theorem, not a postulate. But he was not able to prove it as a consequence of his remaining postulates. The best he could do was to not use it as long as possible. However, he already needed it to prove Theorem I.29, which may be viewed as the beginning of the theory of parallels. It is worth noting that even without the fifth postulate Euclid was able to prove some fairly sophisticated theorems such as the triangle inequality (Theorem I.20). But without the fifth postulate he could not prove, for example, that the sum of three angles of a triangle is always $180°$. For the next two millennia, many unsuccessful attempts were made to prove the fifth postulate (as a consequence of the remaining four). In doing so, it was discovered that the fifth postulate has several equivalent formulations, some of which are:

1. *The sum of three angles of a triangle is $180°$.*

2. Proclus Theorem: *If a line intersects either of two parallel lines, it intersects the other also.*

3. *The opposite sides of a rectangle are equal.*

4. Playfair's Postulate: *Given a line and a point not on the line, there is at most one straight line through the point that does not intersect the given line.*

5. *There is a circle passing through three given non-collinear points.*

6. The Pythagorean theorem.

For Euclid the truth of the fifth postulate was not the issue. The issue was whether it was a theorem, or a necessary postulate. During the 19th century all of a sudden, almost simultaneously, two mathematicians made a

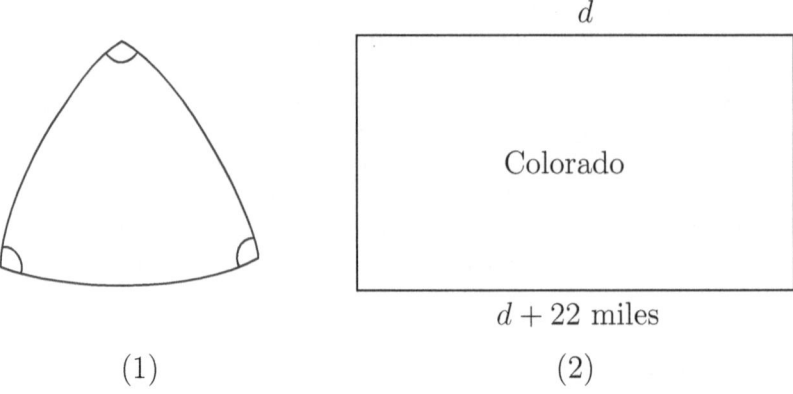

FIGURE 8.10: Figures on curved surfaces.

remarkable breakthrough. They proved the independence of the fifth postulate by constructing geometries without the fifth postulate that turned out to be totally different from that of Euclid's. Such a geometry is called *non-Euclidean geometry*. The mathematicians were **Nikolai Lobachevsky** (1793–1856) from Russia and **János Bolyai** (1802–1860) from Hungary. It is evident from his correspondence that Gauss had also considered non-Euclidean geometry. **Bernhard Riemann** (1826–1866) was a student of Gauss after whom the so-called *Riemannian geometry* is named.

It may be mentioned that in real life the fifth postulate is hard to realize. If a triangle is drawn on a "flat" surface, say on the surface of a calm lake, the sum of three angles will actually be more than 180° as in (1) of Figure 8.10. For another example, look at the map of Colorado, (2) of Figure 8.10 (all of its corners are at right angles). However, the two sides formed by longitudes are not parallel (actually, they meet at the North Pole). Furthermore, its top and bottom sides formed by the latitudes are not equal. This is because the surface of a lake, or that of an ocean, or the earth is only locally flat. Globally the earth is a sphere. See [Gre] for details.

9

Famous Problems from Greek Geometry

Besides demonstrative geometry, such as that in Book I of Euclid, Greeks were also interested in constructive geometry. This required that all constructions in the plane, such as joining two points, or drawing a circle of a given radius and with a given center be done using compass and (uncalibrated) ruler, i.e., a straightedge, only. Some of these constructions are fairly straightforward, while others took 2000 years to be settled, a topic discussed in Chapter 4. Before going any farther we shall need the following theorem, generally attributed to Thales.

- *Any triangle inscribed in a semi-circle is a right triangle.*

Proof. Let ABC be the given triangle. Join C to the center O of the circle. Then we have, by Proposition I.5, the equality of angles as shown in Figure 9.1. So

$$\angle C = \alpha + \beta = \frac{1}{2}(2\alpha + 2\beta) = \frac{1}{2}(\angle A + \angle B + \angle C)$$
$$= \frac{1}{2} \cdot 180° = 90°. \qquad \square$$

Some Basic Constructions

One may wonder why we allow only the straightedge and compass, and no other tool. The answer lies in Euclid's Book I itself. Almost one-third of the propositions in Book I are actually constructions using these two tools. That

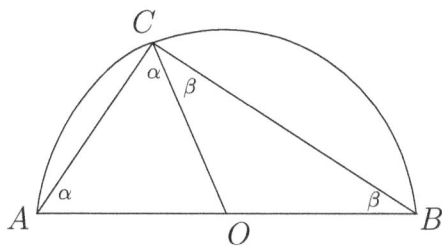

FIGURE 9.1: Thales' theorem.

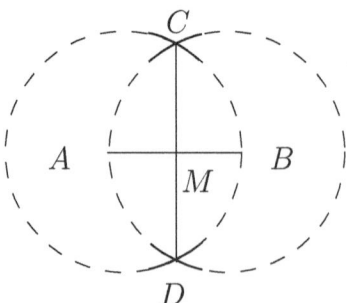

FIGURE 9.2: Bisecting line segments.

we can join two points by a straight line, or extending an existing line, using a straightedge, are essentially Postulates 1 and 2, while Postulate 3 allows drawing circles, using a compass. For example, Proposition 10 shows how to bisect a given line. However, the subtle issue of superposition needs to be pointed out. It is a non-trivial consequence of these three postulates that after opening a compass to a given length, it may be lifted to draw a circle of this radius with any point as its center (see Propositions 2 and 3).

Our first four constructions are basically four propositions in Book I.

(1) *Bisecting a line segment AB.*

Recall Proposition I.1. Open the compass to the length AB. Draw two circles with A and B as centers. Let these two circles intersect at C and D. Join C and D (see Figure 9.2). If CD intersect AB at M, then M is the midpoint of AB. We leave the proof that M is actually the midpoint of AB as an exercise.

(2) *Bisecting an angle.*

Draw a circle with O as center to meet the two lines forming the given angle at A and B as in Figure 9.3.

Draw two more circles of the same radius OA with centers at A and B to intersect at C as shown in Figure 9.3. Join OC. It is easy to see from equal triangles AOC and BOC that OC bisects the given angle.

(3) *Drawing the perpendicular on a line AB from a point P.*

Draw any circle with P as its center to meet the line AB at C and D. From C and D as center draw circles of radius equal to CD to meet at O as shown in Figure 9.4 or Figure 9.5 (depending on whether the point P is on or off the line AB). Join OP. Then $OP \perp AB$.

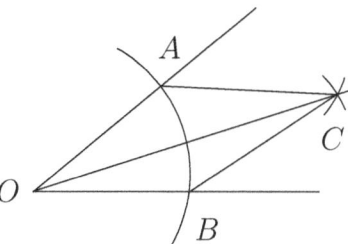

FIGURE 9.3: Bisecting angles.

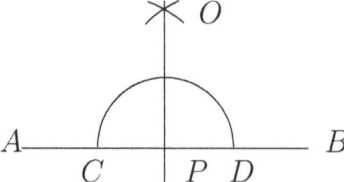

FIGURE 9.4: Perpendicular from a point on a line.

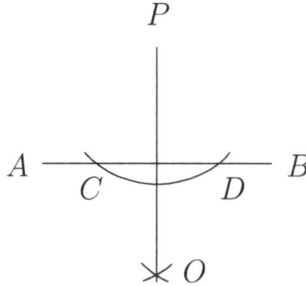

FIGURE 9.5: Perpendicular from points off the line.

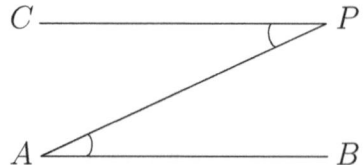

FIGURE 9.6: Drawing a parallel line.

(4) *Drawing a line through a given point P and parallel to a given line AB.*

We may assume that the point P is not on the given line AB for otherwise AB itself is the required line. Join AP and draw angle $CPA =$ angle PAB (see Figure 9.6). That this can be done is left as an exercise. Then $CP \parallel AB$.

(5) *Dividing a line segment AB into n equal parts ($n > 1$ is a positive whole number).*

Draw a line to make an angle with AB and on it mark n points $P_1, \ldots, P_n$ with $AP_1 = P_1P_2 = \cdots = P_{n-1}P_n$ as shown in Figure 9.7. Join P_n to B. From $P_1, \ldots, P_{n-1}$ draw lines parallel to P_nB. This will divide AB into n equal parts as can be seen from similar triangles so formed.

(6) *Starting from unit length constructing a line of length $\sqrt{m}$.*

Here m is a positive integer > 1. To construct a line of length $\sqrt{2}$, construct a right triangle of sides 1 and 1 as shown in Figure 9.8.

Its hypotenuse is of length $\sqrt{2}$. To construct a line of length $\sqrt{3}$, construct a right triangle of sides 1 and $\sqrt{2}$. Its hypotenuse is of length $\sqrt{\sqrt{2}^2 + 1^2} = \sqrt{3}$ and so on. The hypotenuse of a right triangle of sides 1 and $\sqrt{m-1}$ is of length $\sqrt{\sqrt{m-1}^2 + 1^2} = \sqrt{m}$.

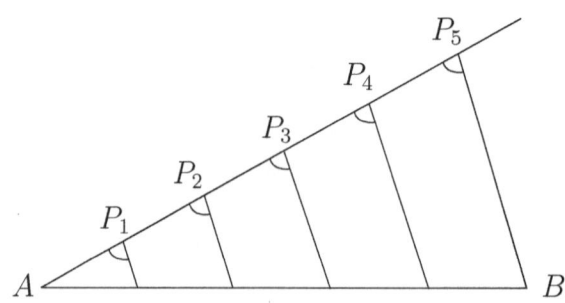

FIGURE 9.7: Dividing line segments into equal parts.

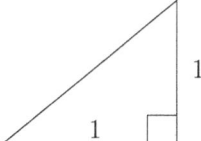

FIGURE 9.8: Constructing an irrational number $\sqrt{m}$.

1. **Exercise.** Suppose m ($m \geq 1$) has no square factor. What is the least number of steps needed to construct $\sqrt{m}$? (A step is drawing a circle or joining two points with a straightedge. Extending an existing line doesn't count.)

(7) *Quadrature of a rectangle.* This means that starting with the sides of a given rectangle, we want to construct a square whose area is equal to that of the rectangle. The side length of this square is called the *geometric mean* of a and b. This is done as follows: Let a and b represent the lengths of the sides of a given rectangle. Surely we can add and subtract lengths, as well as bisect them. So we draw a semicircle of diameter $AB = \frac{a+b}{2}$ as shown in Figure 9.9, and mark C on it so that $AC = \frac{a-b}{2}$. By Thales' theorem, ABC is a right triangle, and by Pythagoras' theorem,

$$BC^2 = \left(\frac{a+b}{2}\right)^2 - \left(\frac{a-b}{2}\right)^2 = ab,$$

and the square constructed on BC will have an area equal to that of the given rectangle of sides a and b.

Corollary. *Quadrature of a triangle ABC.*

Proof. First draw perpendicular CD from C on AB, then bisect CD at M (see Figure 9.10). Let $c = AB$ and $b = MD$. Now construct the square of area $= bc$.

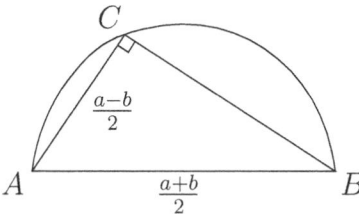

FIGURE 9.9: Quadrature of a rectangle.

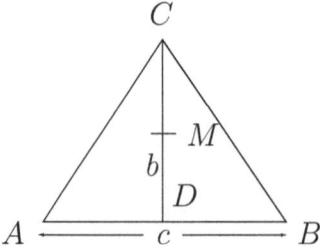

FIGURE 9.10: Quadrature of a triangle.

(8) *Quadrature of the Lune.* A *lune* is a geometric figure, in the plane, bounded by two circles as shown in Figure 9.11, such as the crescent, that is, a new moon. (We regard a circle or a full moon as a lune also.) The Greeks were interested in transforming, using compass and straightedge only, one geometric figure into another of equal area. The squaring or the *quadrature* of lune, or of any other given figure means constructing a square equal in area, using compass and straightedge only. The most outstanding name connected with quadrature of lune is **Hippocrates** of Chios (460–375 BC). Since a circle is a lune, it is possible that Hippocrates was led to the study of lunes by the problem of squaring the circle. He did not square every lune, but a special one as shown in Figure 9.12.

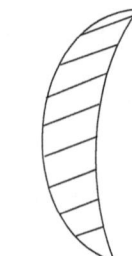

FIGURE 9.11: A lune.

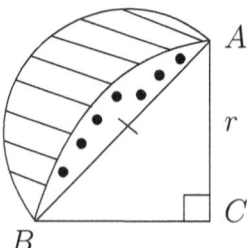

FIGURE 9.12: Quadrature of a lune.

It is the one formed by two related circles, one of radius $BC = r$ say, and the other of radius half of AB, i.e., $\frac{\sqrt{r^2+r^2}}{2} = \frac{r}{\sqrt{2}}$. Hippocrates showed that the area of this lune is the same as that of $\triangle ABC$, not a small feat at his time. In modern notation, area of the lune = area of the smaller semicircle minus the area of the dotted figure. But the area of the dotted figure

$$= \frac{1}{4} \cdot \text{area of bigger circle} - \text{area of } \triangle ABC$$

$$= \frac{\pi r^2}{4} - \frac{r^2}{2}.$$

Hence the area of the lune

$$= \frac{\pi}{2} \cdot \left(\frac{r}{\sqrt{2}}\right)^2 - \left(\frac{\pi r^2}{4} - \frac{r^2}{2}\right)$$

$$= \frac{r^2}{2}$$

$$= \text{area of } \triangle ABC.$$

By the corollary above, we can square $\triangle ABC$, and hence this lune. $\square$

Having done this, the next natural question was how to square a circle. Not only Hippocrates, but many of his successors tried and failed. There are three or four such questions from this period which took mathematicians two millennia to answer.

We will return to them shortly but there is another issue discussed rarely we like to comment on.

Efficiency in Constructions

We now address a related but entirely different question, yet to be addressed by mathematicians. What is the least number of steps needed to carry out a given construction, using only the straightedge and the compass? We must, of course, be precise about what we mean by a step. By definition, a *step* is joining two points by a straightedge, or drawing a (full or part of a) circle. Extending an existing line in either direction is not a step. Choosing a point in the plane is not a step, either.

Recall that to bisect a line segment AB, we need to draw two circles, each of radius AB and with centers at A and B, respectively. We then need the third step to join the two points, common to both the circles, to get the midpoint of AB. It is easy to see that less than three steps will not suffice.

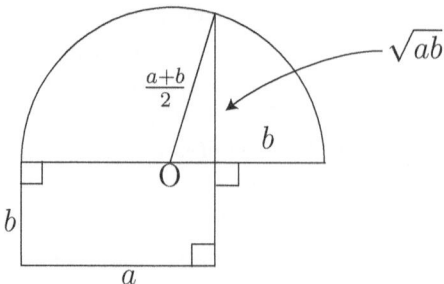

FIGURE 9.13: More efficient quadrature of a rectangle.

Bisection of a given angle takes four steps. We also need four steps to draw $\angle ABC$ on the line segment BC, equal to a given angle. Our method to divide a given line segment AB into n, say $n = 17$, equal parts (see Figure 9.7) took more steps than necessary. Let us count the steps. Choosing a point P_1 not on AB does not count, but joining A to P_1 does. However, extending AP_1 as far as we may need doesn't count as a step. We took sixteen more steps to mark equidistant points $P_2, \ldots, P_{17}$ on the line AP_1. The eighteenth step is to join P_{17} to B. Further, it takes 4 steps to draw a line through P_1 parallel to $P_{17}B$. This produces a line segment of length $\frac{AB}{17}$. Thus the total number of steps is $18 + 4 = 22$.

We leave it as an exercise to figure out if it can be done in fewer than twenty-two steps.

In general, if n is very large, say it has 1 billion binary bits, it is clear from the above discussion that by doubling the distance each time, we need roughly 1 billion (which is $\log_2 n$) steps to mark P_n. In comparison, two, three, or four steps are negligible. So we can mark the nth part of AB in about $\log_2 n$ steps. To calibrate AB with the next $n - 1$ equidistant points takes some n steps. This shows that we can divide the lines AB into n equal parts in roughly $n + \log_2 n$ steps, i.e., in essentially n steps.

For the quadrature of a rectangle, it can be checked that the traditional method (see Figure 9.13) takes fewer steps than the one we gave earlier.

Problems.

In the constructions below, you may use only a straightedge and a compass.

1. What is your least number of steps to divide a given line into fifteen equal parts? Can you prove this number is the smallest possible?

2. Given only the circumference of a circle, locate its center.

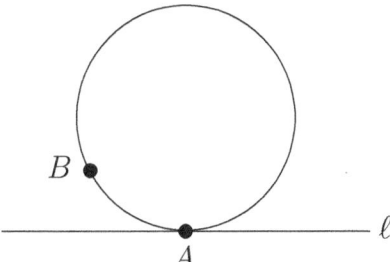

FIGURE 9.14: Circle through a given point and tangent to a given line.

3. Given a circle and a point P outside the circle, draw a line through P and tangent to the circle.

(Note: You may not slide the ruler until it seems to be tangent to the circle.)

4. Given two circles, whose centers are at a distance greater than the sum of their radii, draw tangents common to both circles.

5. Construct the circle inscribed in a given triangle.

6. Construct the circle through the three vertices of a given triangle.

7. Given a line ℓ, a point A on ℓ and a point B not on ℓ (as shown in Figure 9.14), construct the circle through B, which is tangent to ℓ at A.

8. What is the smallest number of steps in which you can construct a square equal to a given triangle? Again, can you ascertain this number of steps is the absolute minimum?

Remark.

The minimum number of steps can be of different magnitude, if we pose the same problem differently. For example, it takes only about $\log_2 n$ steps to construct the nth part of a given line segment AB, whereas to divide AB into n equal parts takes as many as $n + \log_2 n$ steps. Note that $\log_2 n$ is negligible compared to n, if n is extremely large.

Trisection of Angles and Other Famous Problems

Now that we have learned how to divide a line segment into any number of (say three) equal parts, using compass and straightedge alone, it is natural to ask how to divide an angle into three equal parts. The Greeks were also interested in squaring the circle (to construct a square, equal in area to a given circle) and duplicating cubes (constructing the edge of a cube with double the volume of a given cube, hence also called doubling cubes). For a practical aspect of duplicating cubes, see Figure 2.1. On this topic, see the classic book by Felix Klein [Kle]. For two millennia, mathematicians tried in vain to succeed in these tasks. It wasn't until the 19th century that these problems were finally settled.

In 1837, **Pierre Wantzel** (1814–1848) published a proof of the impossibility of two of these tasks. A simple argument based on the roots of polynomials is as follows:

Starting from a line segment of unit length, constructing line segments of other lengths involves intersecting lines and circles. Algebraically, this amounts to successively solving quadratic equations whose coefficients involve the solutions of earlier equations. Thus, the final line segment we are seeking of length α involves square roots of the previously constructed numbers. Hence, the *minimal polynomial* (polynomial of the smallest degree) which α is a root of has degree a power 2^m of 2.

To duplicate a unit cube, we need a line segment of length $\sqrt[3]{2}$. But its minimal polynomial $x^3 - 2$ is not of degree a power of 2. Thus, the duplication of the unit cube is impossible.

The problem of squaring the circle is more subtle. By the same argument as above, first π has to be a root of a polynomial with rational coefficient of degree 2^m. However, in 1882, **Ferdinand Lindemann** (1852–1939) showed that π is transcendental – to begin with, not even a root of any polynomial with rational coefficients. For a proof of this fact, see [Bak-1].

Part III

Contributions of Some Prominent Mathematicians

10

Fibonacci's Time and Legacy

The *Dark Age* is approximately the period AD 500–1250 of European history, whereas the period 1100–1500 was for the awakening there. The millennium 500–1500, between the fall of the Roman Empire in AD 476 and the beginning of the Renaissance, is referred to as the *Middle Ages* or *medieval period*. During the Middle Ages the Catholic Church reigned supreme. The Greek intellectual endeavors were regarded as pagan or evil. The death of Hypatia in 415 had already put an end to the Greek mathematical tradition. Europe plunged into darkness, at least as far as learning and scholarship were concerned. The early Dark Ages had inherited the Greek notion of the quadrivium. A watered-down version of the quadrivium was taught in church schools as "Glory to God," using a text by the Roman scholar **Boethius** (475–524). This text consisted of two books: one consisting of a brief summary of the book on arithmetic from AD 1st century by the Greek mathematician **Nicomachus**, the other containing a few propositions (without proofs) from earlier books of Euclid's *Elements*.

The **Moors** (Muslims from northwestern Africa) entered Spain in 711. While Baghdad was the center of intellectual activity in the East, the cultural centers of the Western Arab world were the Caliphates of Cordova, Seville and Toledo in Spain. In 732 the Arabs were defeated at Tours (in France). Over the next centuries, they were gradually expelled from Spain also. But the Arab tradition of scholarship remained in place in Spain. In 800 the Frankish king Charlemagne (742–824) was crowned by the Pope as the Holy Roman Emperor of the combined territories of France, Germany, Austria and northern Italy. The church debated whether Easter should be determined using the Jewish lunar calendar or the Roman solar calendar. To reconcile the two calendars, a minimal knowledge of mathematics was necessary. Charlemagne recommended that the Easter computations be a part of the curriculum in church schools. He brought **Alcuin of York** (735–804) from England as his educational advisor. Even though this period is called the *Carolingian revival*, the intellectual activity remained low.

The 12th and 13th centuries saw a resurgence of interest in the mathematics from the East. Many important texts on arithmetic, astronomy, algebra, geometry and trigonometry were translated from Arabic into Latin. Even though by this time Spain was a Christian country, it had a thriving Jewish community living in Toledo with many scholars conversant in Spanish as well as in Arabic. First, Arabic texts were translated by Jewish scholars to Spanish and then by Christian scholars from Spanish to Latin. Among the earliest translators were the following:

(1) Adelard of Bath (1075–1164) who was the first to translate Euclid's *Elements*.

(2) John of Seville who translated al Khwarizmi's *Arithmetic* around 1140–1150, whereas al Khwarizmi's *Algebra* was translated in 1145 by Robert of Chester.

(3) The most prolific translator, Gerard of Cremona (1114–1187), was an Italian who worked at Toledo. He translated well over 50 works including classics such as Ptolemy's *Almagest* and Archimedes' *Measurement of Circle*.

During the 13th century the oldest European universities of Bologna, Paris, Oxford and Cambridge were founded.

The most famous or rather the only well-known mathematician of this period was **Leonardo Pisano** (1175–1250), that is, Leonard of Pisa, Italy. He is better known by his nickname **Fibonacci** (filius Bonacci, that is, son of Bonacci). Bonacci was a merchant, and as a youth Fibonacci traveled all over the Mediterranean on behalf of his father. He made several trips to Bugia (now Bejaia in Algeria) where he learned mathematics from the Islamic scholars. He made substantial contributions of his own to what he had learned from the Arabs and wrote three books:

1. *Liber Abaci* (Book of Abacus) in 1202.

2. *Practica Geometriae* (Practice of Geometry) in 1220, and

3. *Liber Quadratorum* (Book on Squares) in 1225.

Liber Abaci

Even though *Liber Abaci* [Fib-1] contained no advance over the mathematics of the Islamic world at that time, this is undoubtedly Fibonacci's most influential book. It introduced and popularized the Hindu numerals in Europe, which the Arabs had learned from the visiting scholars from India. The Hindu numerals had appeared earlier in the work of Gerbert d'Aurillac (945–1003) who later became Pope, but d'Aurillac himself did not understand their full meaning.

It was *Liber Abaci* that taught Europe for the first time how to do arithmetic with the Hindu-Arabic numerals.

Fibonacci began *Liber Abaci* (see [Fib-1]) by introducing these numerals:

"The nine figures of the Indians are 9, 8, 7, 6, 5, 4, 3, 2 and 1. With these nine figures, and with the sign 0, which Arabs call zephirum (cipher) any number can be written as we shall show."

The most famous problem in *Liber Abaci* is the rabbit problem, which leads ultimately to the golden ratio.

The Rabbit Problem: A newly born pair of rabbits of opposite sexes is housed together at the beginning of a year. Starting with the second month, the female gives birth to a pair of rabbits of opposite sexes every month. Each new pair also gives birth to such a pair each month beginning with their second month. What is the number of pairs of rabbits after one year?

If f_n denotes the number of pairs of rabbits at the beginning of the nth month or at the end of $(n-1)$st month, then clearly

$$f_n = f_{n-1} + f_{n-2} \tag{10.1}$$

because all f_{n-1} pairs at month $n-1$ survive to month n and all f_{n-2} pairs from month $n-2$ produce a new pair. The recursive relation (10.1) defines the sequence of the so-called *Fibonacci numbers*

$$0, 1, 1, 2, 3, 5, 8, 13, 21, 34, 55, \ldots$$

These numbers show up everywhere in mathematics.

As f_n is essentially exponential, almost doubling every time, we expect that

$$f_n = q^n \tag{10.2}$$

for some q. But from (10.1) we get

$$q^n = q^{n-1} + q^{n-2}$$

that is

$$q^{n-2}(q^2 - q - 1) = 0.$$

Since $q^{n-2} \neq 0$, we must have

$$q^2 - q - 1 = 0.$$

Hence q is one of the two roots

$$q_1 = \frac{1 + \sqrt{5}}{2} \quad \text{or} \quad q_2 = \frac{1 - \sqrt{5}}{2}$$

of $x^2 - x - 1$. But since q_1 and q_2 are both solutions to the recurrence relation (10.1), which is linear, a general formula for Fibonacci numbers should look like

$$f_n = c_1 q_1^n + c_2 q_2^n.$$

The initial conditions $f_0 = 0$ and $f_1 = 1$ provide the value of c_1 and c_2 giving us an expression for f_n $(n \geq 0)$.

- *The Fibonacci number f_n is given by the formula*

$$f_n = \frac{1}{\sqrt{5}} \left[\left(\frac{1+\sqrt{5}}{2} \right)^n - \left(\frac{1-\sqrt{5}}{2} \right)^n \right].$$

Before proving the assertion, i.e., showing that the above expression for f_n is really the formula for the n-th Fibonacci number, we put (to ease notation)

$$\alpha = \frac{1+\sqrt{5}}{2} \quad \text{and} \quad \beta = \frac{1-\sqrt{5}}{2}.$$

The number α is called the *golden ratio*. The ancient Greeks considered a rectangular figure most pleasing if its edges a and b were in the ratio $a : b = b : a+b$, i.e.,

$$\frac{b}{a} = \frac{a+b}{b}.$$

It is easy to check, first cross-multiplying and then using the quadratic formula, that

$$\frac{b}{a} = \frac{1+\sqrt{5}}{2}.$$

Its conjugate $\beta = \frac{1-\sqrt{5}}{2}$, being negative, cannot be b/a.

The formula is obviously true for $n = 0$ and $n = 1$. To complete the proof by induction, suppose $n \geq 2$ and the formula gives the Fibonacci number for $n - 2$ and $n - 1$. We show that the expression is then also a formula for the n-th Fibonacci number.

Indeed,

$$f_n = f_{n-1} + f_{n-2}$$

$$= \frac{1}{\sqrt{5}} [\alpha^{n-1} - \beta^{n-1}] + \frac{1}{\sqrt{5}} [\alpha^{n-2} - \beta^{n-2}]$$

$$= \frac{1}{\sqrt{5}} [(\alpha^{n-1} + \alpha^{n-2}) - (\beta^{n-1} + \beta^{n-2})]$$

$$= \frac{1}{\sqrt{5}} [\alpha^{n-2}(\alpha + 1) - \beta^{n-2}(\beta + 1)]$$

$$= \frac{1}{\sqrt{5}} [\alpha^{n-2} \cdot \alpha^2 - \beta^{n-2} \cdot \beta^2]$$

$$= \frac{1}{\sqrt{5}} (\alpha^n - \beta^n).$$

Liber Quadratorum

Fibonacci is undoubtedly one of the most influential mathematicians of all times, mainly through *Liber Abaci*. Thanks to *Liber Abaci*, the place value number system devised by Brahmins in India is now used universally throughout the world. Fibonacci numbers, which involve the golden ratio, are familiar to everyone who knows some mathematics. It is, however, his lesser known book, *Liber Quadratorum* (1225) or the Book on Squares [Fib-2]. which contains his original contribution to number theory.

The book, *Arithmeticae* on number theory by the Greek mathematician Diophantus (AD 3rd century) of Alexandria influenced a large number of number theorists, including Fermat. It was in the margin of his copy of *Arithmeticae* that Fermat claimed to have discovered a marvelous proof of Fermat's Last Theorem:

$$x^n + y^n = z^n \quad (n \geq 3) \tag{10.3}$$

has no solution in integers other than the obvious ones. Equations like (10.3) are now called after Diophantus, the Diophantine equations. Solving a Diophantine equation means finding solutions to it in whole numbers, or when equations are homogeneous, equivalently, in rational numbers. For example, the Pythagorean triples are the integer solutions of the Diophantine equation

$$x^2 + y^2 = z^2.$$

Equivalently, they correspond to the points on the unit circle

$$X^2 + Y^2 = 1$$

with rational coordinates $X = \frac{x}{z}$, $Y = \frac{y}{z}$.

The most famous Pythagorean triplet since antiquity is $(3, 4, 5)$. The area of the corresponding right triangle is $\frac{3 \cdot 4}{2} = 6$.

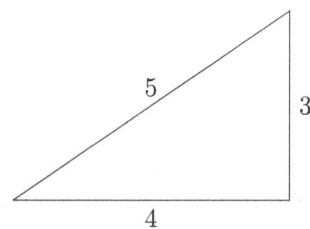

FIGURE 10.1: Pythagorean triplet from antiquity.

In general, n is the area of a right triangle with side lengths α, β, γ if and only if

$$\left.\begin{array}{r} \alpha^2 + \beta^2 = \gamma^2 \\ n = \dfrac{\alpha\beta}{2}. \end{array}\right\} \tag{10.4}$$

Equations (10.6) can be rewritten as

$$\left.\begin{array}{r} \left(\dfrac{\gamma}{2}\right)^2 + n = \left(\dfrac{\alpha+\beta}{2}\right)^2 \\[2mm] \left(\dfrac{\gamma}{2}\right)^2 - n = \left(\dfrac{\gamma-\beta}{2}\right)^2. \end{array}\right\} \tag{10.5}$$

The Persian mathematician **al Karaji** (AD 953–1029) from Karaj near Tehran, was another mathematician influenced by *Arithmeticae* of Diophantus. He asked the following question:

- *For what rational n, is there a square a^2 of a rational number such that $a^2 + n$ and $a^2 - n$ are also squares of rational numbers?*

Looking at (10.4) and (10.5), it is clear that he was asking for rational numbers which are the areas of right triangles with side lengths rational, or right rational triangles, for short. Such numbers are, by definition, *congruent numbers*, although it makes more sense to call them *al-Karaji number*.

If $\triangle_1$ and $\triangle_2$ are similar triangles, then $\triangle_2 = c\,\triangle_1$ where c is the scaling factor. So, the area $|\triangle_2| = c^2|\triangle_1|$. Thus, it suffices to ask: which positive integers $1, 2, 3, 5, 6, 7, \ldots$ with no square factor larger than 1 are congruent numbers?

In the book *Liber Quadratorum* (Book on Squares), Fibonacci discusses this subject and lists two right rational triangles $\left(\frac{3}{2}, \frac{20}{3}, \frac{41}{6}\right)$ and $\left(\frac{35}{12}, \frac{24}{5}, \frac{337}{6}\right)$ with areas 5 and 7, respectively, thus proving that 5 and 7 are congruent numbers. He also stated, without proof, that 1 is not a congruent number. It was Fermat who more than four centuries later, proved that 1, 2, and 3 are not congruent numbers.

Looking at what was known at the time, namely

$$\underbrace{1, 2, 3}_{\substack{\text{non-congruent} \\ \text{numbers}}} \quad ; \quad \underbrace{5, 6, 7}_{\substack{\text{congruent} \\ \text{numbers}}} \;, \ldots$$

a folklore conjecture predicts that

- *every square-free integer $n \equiv 5, 6, 7 \pmod{8}$ is a congruent number.*

In other words, every square-free positive integer n which leaves the remainder 5, 6, or 7 under division by 8 is the area of a right triangle with all side lengths rational numbers.

This is still an unproven conjecture. The most important paper published on this conjecture is by J. Tunnell [Tun]. He proved the validity of this conjecture assuming the unproven conjecture of Birch & Swinnerton-Dyer (see Chapter 15). Like the Riemann hypothesis, the Birch & Swinnerton-Dyer conjecture is one of the most difficult unsolved problems in mathematics from the last millennium, called the *millennium problems*, each carrying a million-dollar prize for the solution.

As mentioned above, Fermat proved that 1, 2, and 3 are not congruent numbers. What about other square-free $n \equiv 1, 2, 3 \pmod 8$?

Given integers $m > 1$ and $0 \le r < m$, all integers $n \equiv r \pmod m$ are the *residue class* of $r \pmod m$. Among all known congruent numbers n (sequence A003273 in OEIS – online encyclopedia of integer sequences), there is no square-free n with $n \equiv 3 \pmod 8$. However, in [Cha-2], the author of this book proved that every residue class $\pmod 8$ contains infinitely many congruent numbers.

Equivalent Formulations of the Congruent Number Problem

We begin with the following:

• *A non-trivial rational solution, i.e., with $Y \ne 0$ of equation*

$$Y^2 = X^3 - n^2 X \tag{10.6}$$

is of the form

$$X = \frac{s}{t^2}, \ Y = \frac{u}{t^3}$$

where the fractions are in the lowest form.

To see this, we write the rational numbers X and Y as $X = s/S, Y = u/U$ with $S \ge 1, U \ge 1$ and GCD $(s, S) = $ GCD $(u, U) = 1$. By plugging in (10.6),

$$u^2 S^3 = U^2 s^3 - n^2 s U^2 S^2.$$

This equation implies that S^3 and U^2 divide each other. Hence $S^3 = U^2$. This can happen only if $S = t^2$ and $U = t^3$ for some $t \ge 1$.

The following is a consequence of this assertion:

- *For a positive integer n, the following are equivalent:*

  (i)   *n is a congruent number;*

  (ii)   *the pair of Diophantine equations*

  $$\left.\begin{aligned} x^2 + ny^2 &= z^2 \\ x^2 - ny^2 &= t^2 \end{aligned}\right\} \tag{10.7}$$

  has a solution in integers with $y > 0$;

  (iii)   *the curve (10.6) has a rational solution with $Y \neq 0$.*

We only need to show that (ii) and (iii) are equivalent.

First suppose equation (10.7) has a solution with $t > 0$. It is easy to see that x, y, z are also non-zero. Multiplying the two equations in (10.7) and dividing throughout by a suitable factor, we obtain

$$\left(\frac{zxt}{y^3}\right)^2 = \left(\frac{x^2}{y^2}\right)^3 - n^2\left(\frac{x^2}{y^2}\right)$$

giving a required solution

$$X = \frac{x^2}{y^2} \quad \text{and} \quad Y = \frac{zxt}{y^3}$$

of (10.6).

Conversely, let $X = s/t^2$, $Y = u/t^3$ be a solution of (10.6). Then

$$\left(\frac{u}{t^3}\right)^2 = \left(\frac{s}{t^2}\right)^3 - n^2\left(\frac{s}{t^2}\right),$$

which gives

$$u^2 = s(s + nt^2)(s - nt^2).$$

A divisibility argument shows that s, $s + nt^2$, $s - nt^2$ are coprime in pairs (have no common factor > 1), hence each is a perfect square. If $s = v^2$, we conclude that

$$v^2 + nt^2 = m^2, \quad v^2 - nt^2 = n^2,$$

and we have produced a required solution of (10.7).

Fibonacci showed that 5 is a congruent number because $41^2 + 5 \cdot 12^2 = 49^2$, $41^2 - 5 \cdot 12^2 = 31^2$.

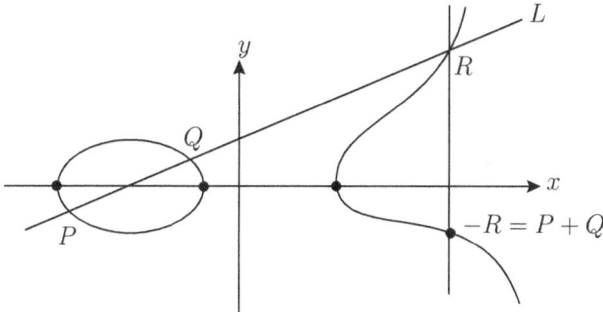

FIGURE 10.2: An elliptic curve.

Elliptic Curves

Recall from Chapter 4 how we transformed equations (4.6) and (4.8) back and forth by rational substitutions. The same is true in general. Any cubic equation with rational coefficients, which defines a smooth curve with a rational point (point with rational coordinates) on it, can be transformed by rational substitutions to

$$y^2 = x^3 + ax + b \qquad (10.8)$$

with a, b in $\mathbb{Z}$ and $4a^3 + 27b^2 \neq 0$. The curve defined by (10.8) is called an *elliptic curve*, and is denoted by E. Pictorially, E looks as in Figure 10.2, assuming $x^3 + ax + y$ has three real roots. Otherwise, it is only the part of it on the right.

Denote by $E(\mathbb{Q})$ the set of points on E, together with the point O at infinity, situated at both ends on the y-axis. Suppose $P = (x_1, y_1)$ and $Q = (x_2, y_2)$ are two points in $E(\mathbb{Q})$. The equation of the chord L joining P and Q (tangent at P, if $P = Q$) is

$$y = mx + \ell, \qquad (10.9)$$

with ℓ and

$$m = \frac{y_2 - y_1}{x_2 - x_1} \qquad (10.10)$$

both rational. Substituting y from (10.9) in (10.8), we get

$$x^3 - m^2 x^2 - (a - 2m\ell)x + b - \ell^2 = 0. \qquad (10.11)$$

If $R = (x_3, y_3)$ is the third point of intersection of the line L with E, then

$$(x - x_1)(x - x_2)(x - x_3) = 0,$$

i.e.,

$$x^3 - (x_1 + x_2 + x_3)x^2 + (x_1 x_2 + x_2 x_3 + x_3 x_1)x - x_1 x_2 x_3 = 0. \qquad (10.12)$$

Comparing the coefficients of x^2 in (10.11) and (10.12), the x-coordinate $x(R)$ of R is given by

$$x(R) = x_3 = m^2 - (x_1 + x_2).$$

The point $-R = (x_3, -y_3)$ is defined to be the sum $P + Q$ of P and Q. Therefore,

$$x(P + Q) = \left(\frac{y_1 - y_2}{x_1 - x_2}\right)^2 - (x_1 + x_2).$$

Now $y(P + Q)$ can be obtained by using equation (10.9).

This short calculation shows that the coordinates of $P+Q$ are also rational.

Note that $R + (-R) = O$, because the chord through R and $-R$ meets the y-axis at the point O infinity. This can also be written as $P + Q + R = O$.

In the language of abstract algebra, $E(\mathbb{Q})$ is an Abelian group with the point O at infinity as its identity element. The study of $E(K)$ with $K = \mathbb{Q}$ and $K = \mathbb{F}_p$, the finite field of p elements (p prime) is one of the most active areas of research in number theory, in fact, in all of mathematics. By the Lutz-Nagell theorem for $E(\mathbb{Q})$, if equation (10.8) has one solution with $t > 1$, it has infinitely many. Consequently, if n is a congruent number, it is the area of infinitely many dissimilar rational right triangles. This is nothing but a tribute to al Karaji and Fibonacci for initiating this subject: very few mathematicians are aware of.

Problems. It is known that 14, 15, and 21 are congruent numbers. Find rational right triangles with areas 14, 15, and 21, respectively.

11

Cardano, del Ferro, and Tartaglia: Solution of the Cubic

Cardano's formula for solving a cubic is the crowning achievement of renaissance mathematics. Yet, it does not receive the same recognition in our curricula as does the quadratic formula, which was discovered long before it. It is rather surprising that there have not been attempts to simplify the messy formulas of Cardano (see [Dun, pp. 611–616]) to a form that would be easier for the students to remember. Apparently, the messy nature of the formulas for solving the cubic is a reason for their lack of popularity. Another reason could be the Galois theory, which modern authors use in their exposition of Cardano's formula. We show that a simple trick, namely a rescaling of the discriminant, reduces not only the formula to a simpler form, but also its verification to a trivial calculation, with no reference to Galois theory. Although, Galois theory is an indispensable tool in algebra and number theory, it is not necessary to wait until one learns it for Cardano's formula. Cardano's formula can be introduced in a first course on complex numbers.

By the celebrated theorem of Abel-Ruffini, a general equation of degree five or more is not solvable by radicals, whereas a quartic equation can be reduced to a cubic equation. Thus Cardano's formula fills an essential gap in our understanding of the solutions of polynomial equations. The purpose of this chapter is to present a simpler exposition of Cardano's formula than found in the literature and to tell the story behind its discovery in order to put the matter in a proper historical perspective.

History

There are reasons to believe that the Babylonians of 2000 BC were familiar with solving quadratic equations, albeit neglecting the negative solutions. Much later, Brahmagupta (c. AD 628) and then al Khwarizmi (AD 780–850), who learned it from him, described the quadratic formula more or less as we know it today. For more on these mathematicians, see Chapter 14. For a wider perspective, see [Wae-2].

The next step was to solve the cubic. The Arabs and the Chinese worked out special cases of the cubic numerically. But it took almost 1000 years from Brahmagupta's time to find a general solution to the cubic, often attributed to the Italian Cardano. However, the story of its discovery is as dramatic as it can be in the world of mathematics.

To solve any polynomial equation, it suffices to take the leading coefficient equal to 1. The so-called *Viète substitution* $X = x - A/3$ reduces the cubic equation

$$X^3 + AX^2 + BX + C = 0$$

to one of the form

$$x^3 + ax + b = 0.$$

Thus there is no loss of generality in assuming that the general cubic equation is monic (leading coefficient 1) and has no square term.

Hindu, Islamic or even the Italian algebra of Cardano's time was entirely rhetorical. There were no symbols for an unknown or its powers. Everything was communicated in words, and to facilitate memorization, formulas were stated as verses. For example, here is part of a verse (see [Car, p. 36]) for the equation $x^3 + px = q$:

> Squeaxno, adtwix
>
> Noesquax, adsub
>
> Axesquno, subadsub
>
> . . .
>
> . . .

It was only after Cardano had published the solution of the cubic in 1545 that **Francois Viète** (1540–1603) introduced, in his book *The Analytic Art*, our present usage of letters to represent unknown quantities. He used vowels for variables and consonants for constants. However, we owe our tradition of using earlier letters $a, b, c, \ldots$ for constants and the later ones $x, y, z, \ldots$ for variables to **René Descartes** (1596–1650). Viète had no symbol for equality. It was Robert Recorde who introduced the symbol $=$ for equality in 1557. The signs $+$ and $-$ appeared for the first time in Germany at the end of the 15th century as symbols for surplus and deficit in business records. In 1514, the Dutch mathematician Vander Hoecke became the first to use them in algebraic expressions. Thomas Harriot was the first (in 1631) to use a dot for multiplication, and he is also responsible for the inequality signs $<$ and $>$. In the same year (1631) William Oughtred introduced the cross sign $\times$ for multiplication. The square root symbol $\sqrt{}$ was invented by Christoff Rudolff (1510–1558), though some historians dispute this attribution. In 1655, John Wallis was the first to use the symbol ∞ for infinity, probably suggested by the late Roman symbol ∞ for a millennium. For more, see [Caj].

Leibniz used these symbols in his calculus, which was popularized by the Bernoullis. The Bernoulli family had great influence on Euler. Finally, it was Euler who utilized these symbols throughout his writings and made them the language of mathematics. Thus the mathematical symbols, which look very intimidating to some people, are very recent phenomena. But they facilitated great advances in mathematics.

Arabs, and the early Europeans who were to start from where the Arabs had left off, did not consider negative coefficients. Thus there were dozens of cases of the cubic equation to be considered. For example, the so-called *depressed form* alone, with square term absent, was split into three cases:

$$x^3 + px + q = 0, x^3 = px + q \text{ and } x^3 + px = q \tag{11.1}$$

with $p, q > 0$.

It was **Scipione del Ferro** (1465–1526), a professor at the University of Bologna, who was the first to find a method of solving equations like (11.1) sometime around 1510. To ensure priority, modern professors announce results even before they have fully checked their proofs. The most famous example of it is the announcement in 1993 by Sir Andrew Wiles that he had proved Fermat's Last Theorem. But the academic life in 16th century Italy was quite different. There was no tenure. University appointments were mostly temporary, subject to periodic renewal. The most common way for professors to stay in their position was to win public contests. A new contender for their job would exchange with them a list of problems to be solved by the other in a specified amount of time. It was required that a solution to every problem submitted must exist. Sometime later, they would meet each other in a public forum to present their solutions; so it was a good strategy for professors to keep their discoveries secret and use them for these public contests. Professor del Ferro never had the occasion to use his solution for such a contest and just before his death in 1526, secretly passed it on to his student Antonio Fiore, as well as his successor Professor **Annibale della Nave** (1500–1558) at the University of Bologna. Even though they never publicized the solution, the news that someone had found a solution to the cubic started to circulate among Italian mathematicians. Another Italian **Nicolo Tartaglia** (1500–1557) from Brescia boasted to have the solution. This was too much for Fiore to take, so he challenged Tartaglia to a public contest. All of Fiore's word problems required the knowledge of a solution to the cubic equation. Having no solution of the cubic yet, Tartaglia was thus trapped, but during the time set aside he worked day and night and, just before the contest on the night of 12 February 1535, found the solution to the cubic. Having worked out the solution himself, Tartaglia easily defeated Fiore who had inherited the solution from his teacher.

At that time **Girolamo Cardano** (1501–1576) was lecturing in Milan on algebra. When he heard about Tartaglia's solution, he wrote to Tartaglia.

He wanted to see the solution so that it could be included in his lectures on algebra. Tartaglia showed the solution only after extracting an oath from Cardano that it would not be included in Cardano's forthcoming book, even with full credit to Tartaglia. Tartaglia wanted to publish it himself. Cardano kept his promise but, assisted by his brilliant student **Lodovico Ferrari** (1522–1565), started working on the problem himself. Ferrari even managed to solve the fourth-degree equation. But their solutions depended on reducing the problem to the cases solved by Tartaglia.

Tartaglia still had not published anything. Cardano did not want to break his promise to Tartaglia–but felt a need to make the solution available to the public. Meanwhile, after hearing the rumor of the original solution by del Ferro, Cardano and Ferrari visited Professor della Nave in Bologna who graciously let them verify that del Ferro indeed had the solution. Cardano no longer felt an obligation to Tartaglia, as he would only be publishing the same solution found independently some 25 years earlier by a mathematician now deceased. Thus in 1545, Cardano published his most important work, *Ars Magna*, mainly devoted to the solution of the cubic. (See Figure 11.1.) When the book appeared, Tartaglia was furious, even though Cardano had mentioned him as one of the original discoverers of the solution. To recoup his prestige, Tartaglia challenged Ferrari to a public contest, but this time he was defeated. To this day, the method described in *Ars Magna* of solving the cubic equation is called Cardano's Formula. We now explain it from a modern point of view, which unifies all the cases into a single formula. For the original case-by-case discussion, see [Car].

Roots of a Complex Number

When we say a polynomial equation is solvable by radicals, we mean the solutions can be found in terms of expressions involving the four algebraic operations on the coefficients of the polynomial, and extracting their square roots, cube roots, and so on. For an integer $m > 1$, we recall how one extracts all the m-th roots of a non-zero complex number. We need it for the next section. For a lively discussion, see [Maz].

A complex number $z = x + iy \neq 0$ can be represented geometrically as a point in the plane

$$z = re^{i\theta} := r(\cos\theta + i\sin\theta),$$

where its *modulus* is the number $r = \sqrt{x^2 + y^2} > 0$ and its *argument* is the angle $\text{Arg}(z) = \theta = \tan^{-1}\left(\frac{y}{x}\right)$.

A well-known formula of de Moivre one learns in a course of trigonometry, is the following fact:

- *For a rational number n/m,* $(\cos\theta + i\sin\theta)^{n/m} = \cos\frac{n}{m}\theta + i\sin\frac{n}{m}\theta.$

HIERONYMI CAR

DANI, PRÆSTANTISSIMI MATHE

MATICI, PHILOSOPHI, AC MEDICI,

ARTIS MAGNÆ,

SIVE DE REGVLIS ALGEBRAICIS,

Lib. unus. Qui & totius operis de Arithmetica, quod

OPVS PERFECTVM

inícripſit,eſt in ordine Decimus,

Habes in hoc libro, ſtudioſe Lector, Regulas Algebraicas (Itali, de la Coſ ſa uocant) nouis adinuentionibus, ac demonſtrationibus ab Authore ita locupletatas, ut pro pauculis antea uulgd tritis, iam ſeptuaginta euaſerint. Neⱥ cⱥ ſolùm, ubi unus numerus alteri, aut duo uni, uerum etiam, ubi duo duobus, aut tres uni ⱥquales fuerint, nodum explicant. Hunc a ũt librum ideo ſeor ſim edere placuit, ut hoc abſtruſiſsimo, & plané inexhauſto totius Arithmeti cæ theſauro in lucem eruto, & quaſi in theatro quodam omnibus ad ſpectan dum expoſito, Lectores incitarẽtur, ut reliquos Operis Perfecti libros, qui per Tomos edentur, tanto auidius amplectantur, ac minore faſtidio perdiſcant.

FIGURE 11.1: *Ars Magna* of Girolamo Cardano (1545).

Let $\sqrt[m]{r}$ be the positive real m-th root of r and ω be the m-th root of unity given by

$$\omega = \cos\frac{2\pi}{m} + i\sin\frac{2\pi}{m}.$$

If we put

$$\alpha = \sqrt[m]{r}\, e^{i\theta/m},$$

then the m-th roots of z are $\alpha, \omega\alpha, \omega^2\alpha, \ldots, \omega^{m-1}\alpha$ (which are clearly distinct, hence account for all of them). In particular, the ratio of any two of them is an m-th root of unity. For example, the three cube roots of $8i$ are

$$2\left[\cos\left(\frac{\pi}{6} + \frac{2\pi}{3}d\right) + i\sin\left(\frac{\pi}{6} + \frac{2\pi}{3}d\right)\right]$$

for $d = 0, 1, 2$.

Cardano's Formula

Recall that the quadratic equation

$$x^2 + bx + c = 0 \tag{11.2}$$

has two solutions

$$x = \frac{-b \pm \sqrt{\Delta}}{2},$$

where the quantity $\Delta = b^2 - 4c$ is called the *discriminant*. The discriminant *discriminates* the solutions. When there is no discriminant, that is, when $\Delta = 0$, the two roots are equal. In fact, the two roots are equal if and only if $\Delta = 0$.

Problem. Solve the quadratic equation (with complex coefficients)

$$\left(1 - \frac{\sqrt{3}}{4}i\right)z^2 - \sqrt{5}\,z + 1 = 0.$$

To solve the cubic we may assume, as has been said earlier, that a general cubic equation is of the form

$$x^3 + ax + b = 0. \tag{11.3}$$

We can define its *discriminant* D such that no two solutions are equal if and only if $D \neq 0$. We make the formulas for the solutions of (11.3) resemble as much as possible that of $x^2 + bx + c = 0$, which are

$$x_1 = \frac{-b + \sqrt{\Delta}}{2} \quad \text{and} \quad x_2 = \frac{-b - \sqrt{\Delta}}{2}.$$

To do this, we modify the traditional definition $D = -(4a^3 + 27b^2)$ of the discriminant of $x^3 + ax + b$ slightly. Our definition of the *discriminant* Δ of $x^3 + ax + b$ is

$$\Delta = \frac{4a^3 + 27b^2}{27}. \tag{11.4}$$

Let ω_1 and ω_2 be the two primitive cube roots of unity:

$$\omega_j = \cos \frac{2\pi j}{3} + i \sin \frac{2\pi j}{3} \ (j = 1, 2).$$

It is easy to check that for $\omega = \omega_1$ or ω_2, $1 + \omega + \omega^2 = 0$. Moreover,

$$\omega_1^2 = \omega_2 \text{ and } \omega_2^2 = \omega_1. \tag{11.5}$$

Now using (11.4), it is easy to check that

$$\frac{-b + \sqrt{\Delta}}{2} \cdot \frac{-b - \sqrt{\Delta}}{2} = -a^3/27.$$

Choose cube roots

$$\alpha_1 = \sqrt[3]{\frac{-b + \sqrt{\Delta}}{2}} \text{ and } \alpha_2 = \sqrt[3]{\frac{-b - \sqrt{\Delta}}{2}} \tag{11.6}$$

such that

$$\alpha_1 \cdot \alpha_2 = -\frac{a}{3}. \tag{11.7}$$

- (del Ferro-Tartaglia-Cardano) *The three solutions of* $x^3 + ax + b = 0$ *are*

$$\alpha_1 + \alpha_2, \ \omega_1\alpha_1 + \omega_2\alpha_2, \ \omega_2\alpha_1 + \omega_1\alpha_2. \tag{11.8}$$

The equations (11.6), (11.7) *and* (11.8) *taken together are called Cardano's formulas.*

Proof. Plug each number from (11.8) in (11.3) and use (11.5), (11.6) and (11.7). □

Examples.

1. To illustrate Cardano's formula, we take one of the simplest examples, namely

$$x^3 - 1 = 0.$$

Here $a = 0$ and $b = -1$, so $\Delta = 1$. From (11.6), we get $\alpha_1 = 1$ and $\alpha_2 = 0$. It follows from (11.8) that the three roots of $x^3 - 1$ are $1 + 0, \omega_1 \cdot 1 + \omega_2 \cdot 0$, $\omega_2 \cdot 1 + \omega_1 \cdot 0$, that is they are $1, \omega_1, \omega_2$. This agrees with what we already know, that is $1, \omega, \omega^2$ are the three solutions of $x^3 - 1 = (x - 1)(x^2 + x + 1) = 0$, where $\omega = \omega_1$ or ω_2 is a primitive cube root of unity.

2. Take $x^3 - 3x + 2 = 0$. Here $a = -3$, $b = 2$, so $\Delta = 0$. The two cube roots of -1 satisfying (11.7) are $\alpha_1 = \frac{1}{2} + \frac{\sqrt{3}}{2} i$ and $\alpha_2 = \frac{1}{2} - \frac{\sqrt{3}}{2} i$. With $\omega_1 = -\frac{1}{2} + \frac{\sqrt{3}}{2} i$ and $\omega_2 = -\frac{1}{2} - \frac{\sqrt{3}}{2} i$, one can check that $\alpha_1 + \alpha_2, \omega_1\alpha_1 + \omega_2\alpha_2$ and $\omega_2\alpha_1 + \omega_1\alpha_2$ are 1, −2, and −2. Note that −2 repeats because $\Delta = 0$.

Problems.

Use Cardano's formulas to solve the following cubic equations:

1. $x^3 - 2x + 4 = 0$,   2. $x^3 + x^2 - 2x - 1 = 0$.

Remarks.

1. Note that choice of the cube roots α_1 and α_2 is dictated by the proof. However, to find the three solutions of (11.3), it is obvious that any choice of α_1 and α_2 will suffice, because the other choices just permute the three numbers in (11.8).

2. In Chapter 10, we discussed briefly the group structure on the set of points $E(\mathbb{Q})$ of the elliptic curve E defined by the equation

$$y^2 = x^3 + ax + b \tag{11.9}$$

with coordinates in $\mathbb{Q}$, together with the point O at infinity. What made the group law work and is crucial throughout is that a straight line intersects the cubic in exactly three points, counted properly. If $\mathbb{Q}$ is replaced by $\mathbb{C}$, then $E(\mathbb{C})$ is a torus (see Figure 11.2). The group structure on $E(\mathbb{C})$ then corresponds to that on $\mathbb{C}/L$ (see [Cha-1], appendix).

FIGURE 11.2: The torus $E(\mathbb{C})$.

12

Leibniz, Newton, and Calculus

Calculus is the study of limits. The concept of limits was not alien to the ancient civilizations, as most of them knew the formula for the area of a circle as derived in Chapter 6. A spectacular event in the history of mathematics was the discovery by Archimedes that the volume of a solid sphere is two-thirds the volume of the smallest cylinder that surrounds it, and that the surface area of the sphere is also two-thirds of the total surface area of the same cylinder (but without its top and bottom), see Figure 12.1. See [Apo] for how Archimedes discovered it. Archimedes must have spent a good part of his life on this project. His method of exhaustion is a forerunner to the definition of the Riemann integral discussed in this chapter.

There is a lot of controversy about whether Leibniz or Newton invented calculus or whether it was "invented" or "discovered." The answer is, it was neither Leibniz nor Newton who discovered or invented calculus. They proved the Fundamental Theorem of Calculus, which had already been stated by Isaac Barrow. It reduced the task of finding the area under the parabola, the surface area, and volume of a sphere to a child's play. As mentioned above, Archimedes spent a good part of his life on these endeavors. In fact, so much so that he was so proud of this achievement that he wanted a sphere and its enveloping cylinder (Figure 12.1) engraved on his grave, although there were other great accomplishments for which he would be remembered forever.

Long before Newton, **Madhav** (AD 1340–1425) in India had already figured out the power series for sine, cosine, and $\tan^{-1}$. In the West, these are referred to as the Taylor series. And it was **Karl Weierstrass** (1815–1897),

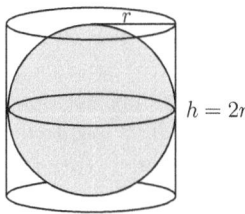

FIGURE 12.1: Sphere in a cylinder.

not Newton, who came up with a precise definition of limit as we know it today, and it goes as follows:

A *metric space* is a non-empty set X of "points" with a notion of distance $d_X(x, y)$ between its points x, y satisfying the usual properties of distance:

(i) $d_X(x, y) = d_X(y, x) \geq 0$, $= 0$ if and only if $x = y$, and

(ii) the triangle inequality–

$$d_X(x, z) \leq d_X(x, y) + d_X(y, z).$$

The real line, Euclidean plane, Euclidean 3-space, and the *Einstein space* of "events" are some examples of *metric spaces*. In the Einstein space, a star exploding at a location (x, y, z) at a time t is an *event* (x, y, z, t). In general, the Euclidean n-space $\mathbb{R}^n$ is the set of points $\boldsymbol{x} = (x_1, \ldots, x_n)$, with coordinates x_j in $\mathbb{R}$, the set of real numbers. The distance between two points $\boldsymbol{x} = (x_1, \ldots, x_n)$ and $\boldsymbol{y} = (y_1, \ldots, y_n)$ in $X = \mathbb{R}^n$ is given by

$$d(\boldsymbol{x}, \boldsymbol{y}) = \sqrt{(x_1 - y_1)^2 + \cdots + (x_n - y_n)^2}.$$

Suppose we are given a function $f : X \to Y$, where (X, d_X) and (Y, d_Y) are metric spaces, and a point a in X. The symbol $\lim_{x \to a} f(x)$ is a point ℓ in Y, if it exists, such that $f(x)$ is arbitrarily close to ℓ, i.e., $d_Y(f(x), \ell)$ is "arbitrarily" small provided the point x in X, $x \neq a$ is "sufficiently" close to a.

The loose terms "arbitrarily" and "sufficiently" can be made precise by using the delta-epsilon definition of Weierstrass, which can be found in any good book on calculus.

A function $f : X \to Y$ is *continuous* at a point a of X if $\lim_{x \to a} f(x)$ exists and is equal to $f(a)$, the "value of f at a."

An illuminating example of a function $f : \mathbb{R} \to \mathbb{R}$ which is discontinuous everywhere is given by the rule

$$f(x) = \begin{cases} 1 & \text{if } x \text{ is rational} \\ -1 & \text{if } x \text{ is irrational.} \end{cases}$$

No matter how close x is to a given a in the domain $X = \mathbb{R}$, there is no ℓ in the codomain $Y = \mathbb{R}$, such that $\text{dist}(f(x), \ell) = |f(x) - \ell|$ is within 1/10th (the chosen closeness) of ℓ. Recall there are rationals and irrationals everywhere, thus $f(x) = 1$ and -1 for x arbitrarily close to a. Therefore, there is no number ℓ such that 1 and -1 are both within 1/10th of ℓ. So, $\lim_{x \to a} f(x)$ does not exist for any a in $X = \mathbb{R}$.

However, most functions dealt with in calculus courses—polynomials, trig, and exponential functions—are continuous. From now on, all functions are assumed to be continuous.

Differential Calculus

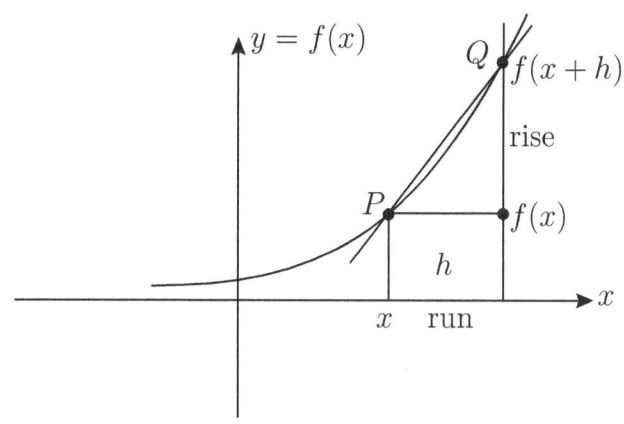

FIGURE 12.2: $\dfrac{\text{rise}}{\text{run}}$.

The *derivative* of a real valued function $f(x)$ of a real variable x is the instantaneous rate of change in $f(x)$ relative to the change in x. In the notation of Leibniz, if an infinitesimal change dx in x produces an infinitesimal change df in f, then the derivative $f'(x) = \frac{df}{dx}$. More precisely, (see Figure 12.2)

$$\frac{df}{dx} = \lim_{h \to 0} \frac{f(x+h) - f(x)}{h}.$$

To stress that h is exceedingly small, usually it is denoted by $\triangle x$.

Examples.

1. $f(x) = c$, a constant. Since there is no rise with run, $\frac{f(x+h)-f(x)}{h} = 0$ for all h. Hence $\frac{df}{dx} = 0$.

2. $y = f(x) = x$.

Everywhere, the ratio $\frac{\text{rise}}{\text{run}} = 1$. Hence $\frac{df}{dx} = 1$.

3. $f(x) = x^2$.

$$\frac{df}{dx} = \lim_{h \to 0} \frac{f(x+h) - f(x)}{h} = \lim_{h \to 0} \frac{(x+h)^2 - x^2}{h}$$

$$= \lim_{h \to 0} (2x + h) = 2x.$$

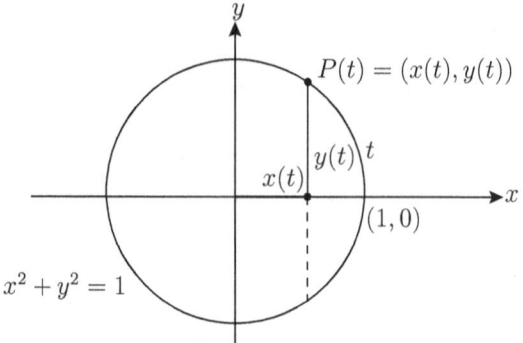

FIGURE 12.3: Definition of $\sin t$ and $\cos t$.

4. In general, if $f(x) = x^n$, a calculation similar to that in Example 3 shows that

$$\frac{d}{dx}(x^n) = nx^{n-1}.$$

Hint: Use the Binomial Theorem for $(x+h)^n$.

5. *Trig Functions.* Whenever asked for the value of $\sin 90$, the students' answer almost always was $\sin 90 = 1$. This is wrong, because $90 \bmod 2\pi$ is not $\pi/2$. So, a clear understanding is necessary of what sine and cosine are as functions of a real variable.

Following **Aryabhata** (AD 476–550) (pronounced Arya-bhatt, not bhataa), $\sin t$ is defined to be the half-chord (see Figure 12.3). More precisely, let $P(t) = (x(t), y(t))$ be the point on the unit circle after traveling distance t on it counterclockwise starting from the point $(1,0)$. (If t is negative, travel clockwise.) Then

$$\left.\begin{array}{l} \cos t = x(t) \\ \sin t = y(t). \end{array}\right\} \tag{12.1}$$

Clearly, $\cos^2 t + \sin^2 t = 1$.

For the interesting etymology of the word sine, see Chapter 14. For its graph, see Figure 12.4.

If t is infinitesimally small, the chord and the arc coincide, and thus $\lim_{t \to 0} \frac{\sin t}{t} = 1$.

The process of computing limits is linear: therefore so is that of derivatives. This means that

(i) $\frac{d}{dx}(f + g) = \frac{df}{dx} + \frac{dg}{dx}$, and

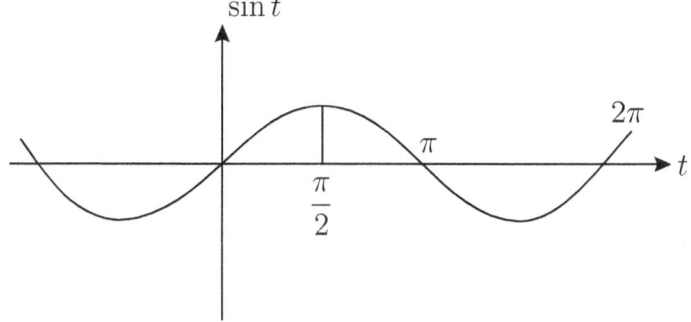

FIGURE 12.4: Graph of sine function.

(ii) for a constant c, $\frac{d}{dx}(cf) = c\frac{df}{dx}$.

An Important Limit

$$\lim_{t\to 0}\frac{\cos t - 1}{t} = \lim_{t\to 0}\frac{\cos t - 1}{t}\cdot\frac{\cos t + 1}{\cos t + 1} = \lim_{t\to 0}\frac{-\sin^2 t}{t(\cos t + 1)}$$

$$= -\lim_{t\to 0}\frac{\sin t}{t}\cdot\frac{\sin t}{\cos t + 1} = 0.$$

In view of the above limit, $\frac{d}{dx}(\sin x) = \lim_{h\to 0}\frac{\sin(x+h)-\sin x}{h}$, which by the trig formula $\sin(x+h) = \sin x\cos h + \sin h\cos x$ is

$$= \lim_{h\to 0}\left[\sin x\,\frac{\cos h - 1}{h} + \cos x\,\frac{\sin h}{h}\right]$$

$$= \cos x.$$

Hence

$$\boxed{\frac{d}{dx}(\sin x) = \cos x.}$$

Similarly, $\frac{d}{dx}(\cos x) = -\sin x$.

Exponential Function

Choose a base $a > 0$, e.g., $a = 2$ or 10. Recall that $a^{-1} = \frac{1}{a}$. If we put $a^m = \underbrace{a\cdots a}_{m-\text{times}}$ and $a^{\frac{1}{n}} = \sqrt[n]{a}$, then clearly, $a^{m+n} = a^m\cdot a^n$, $a^0 = a^{1-1} = \frac{a}{a} = 1$.

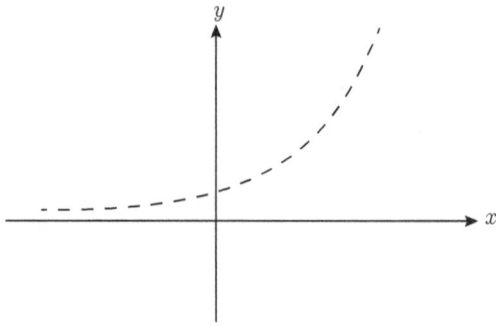

FIGURE 12.5: Exponential function a^r for rational r.

Thus, for a rational $r = \frac{m}{n}$, the function $f(r) = a^r = \sqrt[n]{a^m}$ is well-defined. The graph of this function $f : \mathbb{Q} \to \mathbb{R}$ is shown in Figure 12.5.

It is extended to the function $f : \mathbb{R} \to \mathbb{R}$ by continuity – the graph is unbroken.

Recall the definition of a bijective map, or function $f : X \to Y$. For any given y in Y, there is one and only one x in X with $y = f(x)$. This defines the inverse f^{-1} of f by setting $f^{-1}(y) = x$. For example, if $\mathbb{R}^+$ is the set of positive reals, the inverse of the squaring function $s : \mathbb{R}^+ \to \mathbb{R}^+$ given by $s(x) = x^2$ is the square root function $\mathrm{sqrt}(x) = \sqrt{x}$.

Since the exponential function $f : \mathbb{R} \to \mathbb{R}^+$ given by $y = f(x) = a^x$ is bijective, by definition, the logarithm in base a is its inverse $\log_a : \mathbb{R}^+ \to \mathbb{R}$, i.e., $\ln_a(x) = y$. For example, since $32 = 2^5$, $\log_2(32) = 5$. Since $a^{x+y} = a^x \cdot a^y$ and $a^{x-y} = \frac{a^x}{a^y}$,

$$\left. \begin{array}{c} \log_a(xy) = \log_a(x) + \log_a(y) \\[2mm] \log_a\left(\frac{x}{y}\right) = \log_a(x) - \log_a(y) \\[4mm] c\log_a x = \log_a x^c. \end{array} \right\} \qquad (12.2)$$

Also, clearly

There is a particular base denoted by e which is the most natural base for the reason to be given shortly. Most often, it is introduced in a clumsy way, e.g., a book on calculus says, "among all the bases, there is one which outshines all others. It is denoted by the letter e. Roughly, $e = 2.718$."

One should wonder, what is wrong with $a = 10$? Why doesn't it outshine others? After all, it is built into our daily life.

To explain what exactly e is and why it is natural, suppose we borrow \$1 for one year at an annual rate of $1/1$ (i.e., 100%). If the interest is simple (not compounded), the amount owed at the end of the year is \$2. However, if it is compounded biannually, the amount owed after six months is $1 + \frac{1}{2}$. During the next six months, the interest is applied to \$1.5. Hence, the amount owed now is

$$\left(1 + \frac{1}{2}\right)\left(1 + \frac{1}{2}\right) = \left(1 + \frac{1}{2}\right)^2.$$

Similarly, if the interest is compounded quarterly, expect it to be

$$\left(1 + \frac{1}{4}\right)^4.$$

Usually, the interest is compounded daily. So, the amount owed at the end of the year is

$$\left(1 + \frac{1}{365}\right)^{365}.$$

However, the interest should be compounded continuously and this figure should be

$$\lim_{n \to \infty} \left(1 + \frac{1}{n}\right)^n.$$

The *Euler number* is this limit and is denoted by e. An approximate value of $e = 2.718$.

In general, if a dollar is borrowed for t years at the annual rate $r/$unit, the above figure, say $P(t)$, is given by

$$P(t) = \lim_{n \to \infty} \left[\left(1 + \frac{rt}{n}\right)^{\frac{n}{rt}}\right]^{rt}$$

$$= \lim_{N \to \infty} \left[\left(1 + \frac{1}{N}\right)^N\right]^{rt}$$

$$= e^{rt}.$$

So,

$$\boxed{P(t) = e^{rt}.}$$

This is the exponential growth in say, bacteria.

Derivative of Logarithm Function

When the base e is used for the exponential function, sometimes we write $e^x = \exp(x)$, in which case, $\exp^{-1} = \ln$, the natural logarithm.

To compute $\frac{d}{dx}(\ln x)$, by (12.2),

$$\frac{\ln(x+h) - \ln x}{h} = \frac{1}{h} \cdot \ln\left(1 + \frac{h}{x}\right)$$

$$= \ln\left(1 + \frac{h}{x}\right)^{\frac{1}{h}}$$

$$= \ln\left[\left(1 + \frac{h}{x}\right)^{\frac{x}{h}}\right]^{\frac{1}{x}}.$$

Since $x > 0$ is fixed,

$$\frac{d}{dx}(\ln x) = \lim_{h \to 0} \frac{\ln(x+h) - \ln x}{h}$$

$$= \ln e^{\frac{1}{x}} = \frac{1}{x}$$

i.e.,

$$\boxed{\frac{d}{dx}(\ln x) = \frac{1}{x}.}$$

Chain Rule

One of the most useful tools in calculus is the Chain Rule for the so-called "function of a function." If $f(u)$ is a function of u and $u(x)$ is a function of x, then $f(u(x))$ is a function of a function. The Chain Rule asserts that

$$\frac{df(u(x))}{dx} = \frac{df}{du} \cdot \frac{du}{dx}.$$

Our notation contains the hint for its obvious proof.

An Application

Suppose $u(x) = e^x$, $f(u) = \ln u$. Then $f(u(x)) = \ln e^x = x$. Hence,

$$\frac{df}{dx} = \frac{df}{du} \cdot \frac{du}{dx} = \frac{1}{u} \cdot \frac{d}{dx}(e^x).$$

On the other hand, $\frac{d}{dx}(x) = 1$. Thus,

$$\frac{d}{dx} e^x = u = e^x$$

$$\boxed{\frac{d}{dx} e^x = e^x.}$$

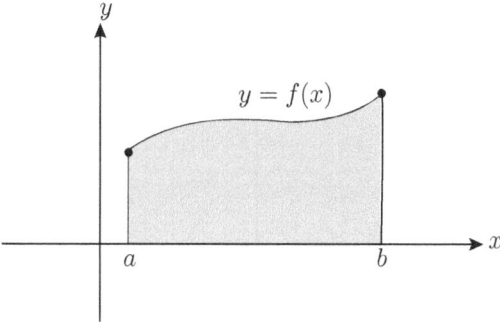

FIGURE 12.6: Definite integral.

Integral Calculus

The method of exhaustion Archimedes used to find the volume and the surface area of a sphere, a simpler example of which is the derivation of the area of a circle in Chapter 6, is a precursor of the so-called Riemann integral.

Suppose $f : [a, b] \to \mathbb{R}$ is a continuous function with domain a closed interval $I = [a, b]$ of length $\ell|I| = b - a$. The *definite integral*

$$\int_a^b f(x)dx$$

is by definition the area under the graph of $y = f(x)$ from $x = a$ to $x = b$ (see Figure 12.6).

It can be approximated to any degree of accuracy by the so-called Riemann sums, to define which we divide the interval I into n subintervals I_j, each of length $\ell|I|/n$. Then, choose a point a_j in each I_j. The *Riemann sum*, which of course depends on the choice of a_j, is the following approximation (see Figure 12.7) to the area:

$$R_n(f) = f(a_1)\ell|I_1| + \cdots + f(a_n)\ell|I_n|.$$

Thus,

$$\int_a^b f(x)dx = \lim_{n \to \infty} R_n(f),$$

a definition by the method of exhaustion.

Given arbitrarily chosen functions, say $f(x) = \sin x$, $\ln x$ or e^x, even Archimedes could not compute the area

$$\int_a^b f(x)dx.$$

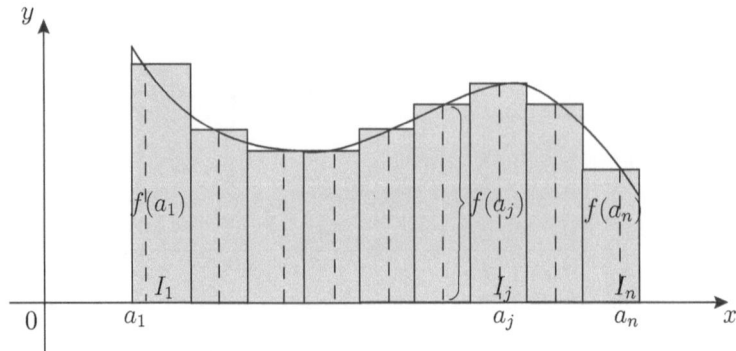

FIGURE 12.7: Riemann sum (shaded area).

The most spectacular discovery in the history of mathematics, with almost a trivial proof, was that of the Fundamental Theorem of Calculus (FTC). It is so powerful that the calculations on which Archimedes spent a good part of his life are now almost a trivial task. Its first published statement by **Isaac Barrow** (1630–1677), a predecessor of **Isaac Newton** (1643–1727) at the University of Cambridge, can be found in his lecture notes of 1674. Isaac Newton, **Gottfried Leibniz** (1646–1716), and **James Gregory** (1638–1675) are all credited with its proof, which is a triviality compared to the proof by Gauss of his theorem on the constructibility of regular n-gons.

To prepare for the statement and proof of FTC, we begin with the following:

A function $F(x)$ is a *differentiable function* if its derivative $\frac{dF}{dx} = F'(x)$ exists. An *antiderivative* of a continuous function $f(x)$ is a differentiable function $F(x)$ with $F'(x) = f(x)$. The traditional notation for $F(x)$ is $\int f(x)dx$ (without limits on $\int$) and is called the *indefinite integral*, whereas the area $\int_a^b f(x)dx$ is called the *definite integral*. Without going into the etymology of these terms, which is rooted in the history of FTC, we now state and prove the

Fundamental Theorem of Calculus

• *A continuous function $f(x)$ with its domain a closed interval $[a,b]$ has an antiderivative $F(x)$, and the area*

$$\int_a^b f(x)dx = F(b) - F(a) = F(x)|_a^b.$$

The proof below of the FTC contains no hint on how to find $F(x)$ in a "closed form" for an arbitrarily given $f(x)$, unless $f(x)$ happens to be the

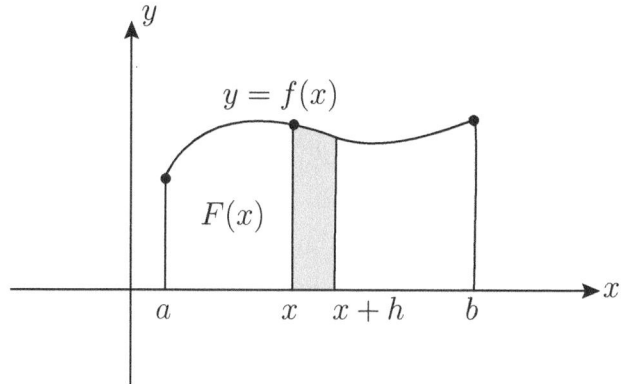

FIGURE 12.8: Proof of FTC.

derivative of some standard functions whose derivatives are already known to us (see Table 12.1).

Proof of FTC

Set

$$F(x) = \int_a^x f(t)dt,$$

the area under the graph of $y = f(x)$ from $t = a$ to $t = x$ (see Figure 12.8).

Then $F(x + h) - F(x)$ is the area of the shaded region and for h exceedingly small, it is almost equal to $hf(x)$. Therefore,

$$\frac{F(x + h) - F(x)}{h} = f(x) + \epsilon(h)$$

where the error $\epsilon(h)$ disappears as $h \to 0$. Thus,

$$F'(x) = \lim_{h \to 0} (f(x) + \epsilon(h))$$

$$= f(x).$$

Since by definition of $F(x)$,

$$\int_a^b f(x)dx = F(b) \text{ and } F(0) = 0,$$

we have

$$\int_a^b f(x)dx = F(b) - F(a). \qquad \square$$

TABLE 12.1: Table of antiderivatives.

$F(x)$	$f(x)$
$\dfrac{x^{n+1}}{n+1}$	x^n
$\sin x$	$\cos x$
$-\cos x$	$\sin x$
$\ln x$	$\dfrac{1}{x}$
e^x	e^x

Application of FTC

1. Archimedes spent a good amount of time on figuring out [Dun] the area A enclosed by a parabola (quadrature of a parabola) (see Figure 12.9).

   The FTC reduces it to a trivial calculation. For $F(x) = \frac{x^3}{3}$, $F'(x) = x^2$. Hence, the area under $y = x^2$ for $x = 0$ to $x = a$ is $= \int_0^a x^2 dx = F(a) - F(0) = \frac{a^3}{3}$. Thus, $A = 2\left(a^3 - \frac{a^3}{3}\right) = \frac{4a^3}{3}$.

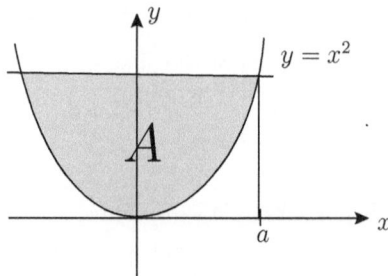

FIGURE 12.9: Quadrature of a parabola.

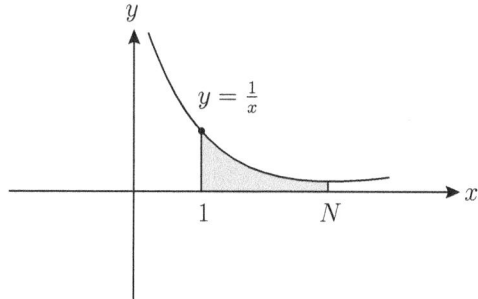

FIGURE 12.10: Area under $y = \frac{1}{x}$.

2. Area under the graph of $y = \frac{1}{x}$ for $x = 1$ to $x = N$ (see Figure 12.10) is

$$\int_1^N \frac{1}{x}\,dx = \ln x|_1^N$$
$$= \ln N - \ln 1$$
$$= \ln N, \text{ since } \ln 1 = 0.$$

This will be needed in the next chapter.

3. The area enclosed by one arc of $y = \sin x$ (see Figure 12.11) is

$$\int_0^\pi \sin x\,dx = -\cos x|_0^\pi = -(\cos \pi - \cos 0) = 2.$$

4. We leave it for the reader to show that the surface area of a sphere of radius a is $4\pi a^2$.

5. *Volume of sphere* of radius $= a$.

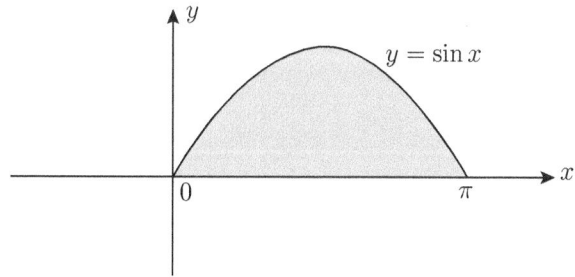

FIGURE 12.11: Arc of sine function.

Let S_r be the concentric shells of radii r and thickness dr. The volume of S_r is, by the previous example, $4\pi r^2 dr$. Hence, the required volume is

$$4\pi \int_0^a r^2 \, dr = \frac{4\pi}{3} a^3.$$

13

Euler and Modern Mathematics

The 18th century produced some of the greatest mathematicians of all time, such as **Jean-Joseph Fourier** (1768–1830), **Joseph Louis Lagrange** (1736–1813), Pierre Laplace (1749–1827), and **Adrien Marie Legendre** (1752–1883), but Leonhard Euler (1707–1783) had no rival. He is the one who brought mathematics to its current form. He introduced notations which are now standard in mathematics. Among numerous notations, we mention just a few: π for universal ratio of the perimeter of a circle to its diameter, i for $\sqrt{-1}$, e for the natural base for the exponential function, $f(x)$ for a function of x, the notation $\cos t + i \sin t = e^{it}$ to define the exponential function of a complex variable, which led to the so-called most famous equation $e^{\pi i} + 1 = 0$ in mathematics. He used Σ for sum and introduced the convention to label the side lengths of a triangle opposite to angle A, B, C with lowercase a, b, c.

In number theory, he introduced the *Euler ϕ-function* $\phi(n)$ as well as the gamma constant. To give a new proof of the infinitude of prime numbers, he introduced the zeta function of a real variable, which **Bernhard Riemann** (1826–1866) extended to the Riemann zeta function of a complex variable to count the number of primes $p \leq x$, where x is an arbitrarily large number. To do this, he stated the Riemann hypothesis, which some regard as the most challenging problem in mathematics. Euler also generalized the factorial function $n! = 1 \cdot 2 \cdot 3 \cdots n$ to the gamma function $\Gamma(s)$, defined by

$$\Gamma(s) = \int_0^\infty e^{s-1} e^{-t} dt,$$

for $s = \sigma + it$, $\sigma > 0$. Note that $\Gamma(n) = (n-1)!$

There is not a field of mathematics, pure or applied, to which he did not contribute substantially. Not only that, he also created several brand new subjects of mathematics such as analytical number theory, graph theory, and topology, to mention a few. While Gauss is called the Prince of Mathematics, the title of King of Mathematics is generally awarded to Euler. His work, which is still under compilation, already consists of ninety volumes, enough to fill the whole shelf in a library. It is agreed, generally, that Euler was the greatest mathematician of all time.

Euler was born in 1707 in Basel, Switzerland. He was tutored privately by Johann Bernoulli. At the age of 20, he followed Johann's son Daniel Bernoulli to St. Petersburg, Russia. Daniel was going to hold a chair in mathematics at the Russian Academy, where he secured a position for Euler also, but only in medicine. In 1733, Daniel left Russia, and Euler moved into his chair. In the mid-1730s, Euler lost his right eye. In 1741, he also left Russia for the Berlin Academy under Frederick the Great, but returned to Russia at the invitation of Catherine the Great and stayed there for 17 years until his death in 1783.

Throughout his life, Euler was blessed with an extraordinary memory. He memorized not only the first one hundred primes but also their squares, cubes, as well as their fourth, fifth, and sixth powers. Once he mentally summed seventeen terms of a complicated series with fifty place accuracy. In 1771, he became almost totally blind before he was 60 years old but could dictate his papers and books to his attendants and some of his thirteen children. Both in quality and quantity, Euler's achievements are unsurpassed. He wrote excellent texts on algebra, analysis, and other subjects. The texts on basic algebra in American colleges are modeled after Euler's *Algebra*. This book on algebra has been a standard text for two centuries. In this chapter, we will discuss only–and briefly–his work in number theory, graph theory, and topology. For a short commentary on his work, see [Bak-2].

Number Theory

Fermat, considered to be the father of modern number theory, had made two statements without proof:

1. *Every prime $p \equiv 1$* (mod 4) *is a unique sum*

$$p = a^2 + b^2$$

of two squares.

2. *For every $n \geq 0$, $F(n) = 2^{2^n} + 1$ is prime.*

Euler supplied a proof of statement 1, and disproved statement 2 in two different ways. Fermat had already verified that statement 2 is true for $n = 0, 1, 2, 3$, and 4, but $F(5) = 4,294,967,297$ was too large for Fermat to check. However, Euler showed that $F(5)$ is divisible by 651, thus not a prime. He also showed that

$$F(5) = (2^{16})^2 + 1$$

$$= 62264^2 + 20449^2,$$

contradicting the uniqueness in statement 1.

This reminds us of Ramanujan's reply to Hardy's remark that the taxi he took to visit him had a rather dull number, 1729. At the spur of the moment,

Ramanujan said, "No, it is a very interesting number. It is the smallest number that can be written as a sum of two cubes in two different ways: $1729 = 1^3 + 12^3 = 9^3 + 10^3$."

Euler also gave a proof of another statement of Fermat, known as *Fermat's Little Theorem.*

- *If p is a prime, then for every natural number a,*

$$a^p \equiv a \pmod{p}.$$

(In the language of finite fields, it says that the multiplicative order of a non-zero element of a finite field $\mathbb{F}_p$ of p elements is a factor of $p - 1$.)

Moreover, he generalized it to the following:

- *If $m > 1$ is any modulus and a is an integer with* GCD. $(a, m) = 1$, *then*

$$a^{\phi(m)} \equiv 1 \pmod{m},$$

where $\phi(m)$ is the Euler ϕ-function.

In cryptography the RSA, used for secure communications, is based on this theorem of Euler.

Analytical Number Theory

The harmonic series

$$1 + \frac{1}{2} + \frac{1}{3} + \frac{1}{4} + \frac{1}{5} + \cdots$$

which Euler denoted by

$$\sum_{n=1}^{\infty} \frac{1}{n}$$

diverges. This means the sum of enough number of terms exceeds any given number, however large.

This is so because the partial sums,

$$S_N = \sum_{n=1}^{N} \frac{1}{n}$$

$$= 1 + \frac{1}{2} + \frac{1}{3} + \cdots + \frac{1}{N}$$

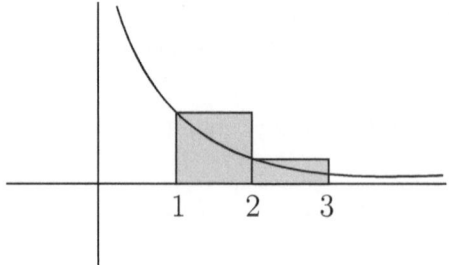

FIGURE 13.1: Convergence of harmonic series.

are larger than the area under the curve $y = \frac{1}{x}$ from $x = 1$ to $x = N$ (see Figure 13.1), in other words,

$$S_N \geq \int_1^N \frac{1}{x}\, dx = \ln x_{|1}^N = \ln N,$$

which increases with no upper bound.

Euler's idea to prove infinitude of primes was to put $u = \frac{1}{p}$ in

$$\frac{1}{1-u} = 1 + u + u^2 + u^3 + \cdots$$

so that

$$\left(1 - \frac{1}{p}\right)^{-1} = 1 + \frac{1}{p} + \frac{1}{p^2} + \frac{1}{p^3} + \cdots.$$

Taking infinite product over all primes p,

$$\prod_p \left(1 - \frac{1}{p}\right)^{-1} = \sum \frac{1}{p_1^{e_1} \cdots p_r^{e_r}}.$$

In the last sum, each product $p_1^{e_1} \cdots p_r^{e_r}$ $(e_j \geq 0)$ appears exactly once. Hence, by the Fundamental Theorem of Arithmetic,

$$\sum \frac{1}{p_1^{e_1} \cdots p_r^{e_r}} = \sum_{n=1}^{\infty} \frac{1}{n}.$$

To summarize, the Fundamental Theorem of Arithmetic is encoded in the following, the so-called Euler Product.

Euler Product

$$\prod_p \left(1 - \frac{1}{p}\right)^{-1} = \sum_{n=1}^{\infty} \frac{1}{n}. \tag{13.1}$$

To prove the infinitude of primes, suppose there are only finitely many primes. Then the LHS of equation (13.1) is a finite number, whereas its RHS is infinite, which is not possible. □

Actually, Euler defined the zeta function of a real variable by the infinite series

$$\zeta(\sigma) = \sum_{n=1}^{\infty} \frac{1}{n^\sigma} = 1 + \frac{1}{2^\sigma} + \frac{1}{3^\sigma} + \frac{1}{4^\sigma} + \cdots . \tag{13.2}$$

The series on the right of (13.2) converges for $\sigma > 1$, by a similar argument used to prove the divergence of the harmonic series

$$\sum_{n=1}^{\infty} \frac{1}{n} = 1 + \frac{1}{2} + \frac{1}{3} + \frac{1}{4} + \cdots .$$

Euler computed $\zeta(\sigma)$ for even values of $\sigma = 2m$ with $m = 1, 2, 3, \ldots, 13$, and so on.

$$\zeta(2) = 1 + \frac{1}{2^2} + \frac{1}{3^2} + \frac{1}{4^2} + \cdots = \frac{\pi^2}{6}$$

$$\zeta(4) = 1 + \frac{1}{2^4} + \frac{1}{3^4} + \frac{1}{4^4} + \cdots = \frac{\pi^4}{90}$$

$$\zeta(6) = 1 + \frac{1}{2^6} + \frac{1}{3^6} + \frac{1}{4^6} + \cdots = \frac{\pi^6}{945}$$

$$\vdots$$

$$\zeta(26) = 1 + \frac{1}{2^{26}} + \frac{1}{3^{26}} + \frac{1}{4^{26}} + \cdots = \frac{1315862}{11094481976030578125} \pi^{26} .$$

In general, for $m \geq 1$,

$$\zeta(2m) = \frac{2^{m-1} B_m}{(2m)!} \pi^{2m}, \tag{13.3}$$

where B_m are the Bernoulli numbers which occur in the Taylor expansion:

$$\frac{x}{e^x - 1} = 1 - \frac{x}{2} + \sum_{m=2}^{\infty} (-1)^{m-1} \frac{B_m}{(2m)!} x^{2m} .$$

The first few Bernoulli numbers B_j for $j = 0, 1, 2, 4, 6, 8, 10, \ldots$ (others with odd j are zero) are $1, -1/2, 1/6, -1/30, 1/42, -1/30, 5/66, \ldots$.

Euler's Discovery of $e^{\pi i} + 1 = 0$

How did Euler discover this equation? He preferred to think of transcendental functions in terms of their Taylor series, e.g.,

$$e^x = 1 + x + \frac{x^2}{2!} + \frac{x^3}{3!} + \cdots$$

$$\cos x = 1 - \frac{x^2}{2!} + \frac{x^4}{4!} - \frac{x^6}{6!} + \cdots$$

$$\sin x = x - \frac{x^3}{3!} + \frac{x^5}{5!} - \frac{x^7}{7!} + \cdots .$$

This led him to the correct definition of the complex exponentiation e^z for $z = s + it$. Since the correct definition must have the property: $e^{s+it} = e^s \cdot e^{it}$, it is enough to define e^{it}. Thus, thinking of e^{it} as a series, just like one for e^x,

$$e^{it} = 1 + it + \frac{(it)^2}{2!} + \frac{(it)^3}{3!} + \frac{(it)^4}{4!} + \cdots$$

$$= 1 - \frac{t^2}{2!} + \frac{t^4}{4!} - \cdots$$

$$+ i \left(t - \frac{t^3}{3!} + \frac{t^5}{5!} - \cdots \right)$$

$$= \cos t + i \sin t$$

- $\boxed{e^{it} = \cos t + i \sin t.}$

If we plug $t = \pi$ in it, which is actually not an equation, but the definition of e^{it}, we get

$$\boxed{e^{\pi i} + 1 = 0.}$$

It is claimed by many to be the most important equation in mathematics, as it ties the most important five numbers, $0, 1, e, \pi, i$ in a single equation. The most quoted admirer of this equation seems to be the Nobel Laureate physicist Richard Feynman.

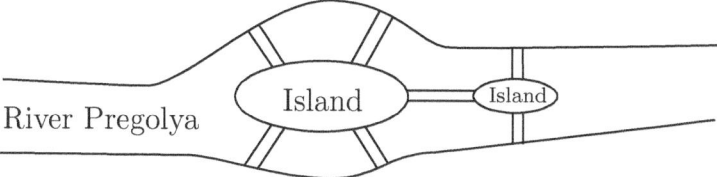

FIGURE 13.2: Seven bridges of Königsberg.

Graph Theory and Topology

Graph theory and topology were born in 1735 with Euler's solution of the Königsberg bridge problem. A recreational puzzle, it asks for a route, starting and ending at the same place, which crosses each of the seven bridges in Königsberg (now Kaliningrad, Russia) exactly once (see Figure 13.2).

To solve this problem, Euler created what are now called graph theory and topology. He represented land masses and islands by a set of four points or vertices and the bridges by lines, called edges, between them (see Figure 13.3).

This is an example of a *multigraph* where two or more edges are allowed between two vertices. Another example of a multigraph is the molecular diagram for the benzene molecule (see Figure 13.4). A graph is a *simple graph* if between given two vertices, at most one edge is allowed.

The *degree of a vertex v*, denoted by $\deg(v)$ is the number of edges going out of it. Thus, the *degree sequence* of the graph representing Königsberg bridges is $5, 3, 3, 3$. In his honor, a graph is a *Eulerian graph* if starting and ending at the same vertex, each edge can be traversed exactly once.

Euler proved one of the first theorems in graph theory, which we state now:

- *A graph is Eulerian if and only if all its vertices have even degrees.*

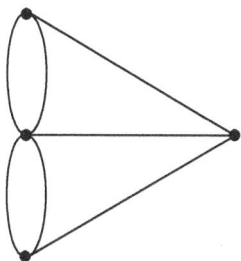

FIGURE 13.3: Graphical representation of Königsberg bridges.

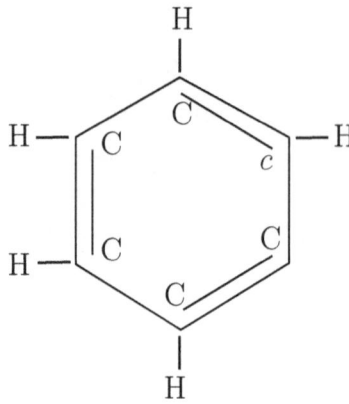

FIGURE 13.4: Benzene molecule.

Thus, the Königsberg bridge problem has no solution.

A class of graphs which has been studied more widely is that of simple graphs, which we make precise by the following definition:

If V is a non-empty set, the set of its subsets each having cardinality r is denoted by $\mathcal{P}_r(V)$. For example, if $V = \{1, 2, 3\}$, then

$$\mathcal{P}_1(V) = \{\{1\}, \{2\}, \{3\}\}, \ \mathcal{P}_2(V) = \{\{2, 3\}, \{3, 1\}, \{1, 2\}\},$$

and $\mathcal{P}_3(V) = \{\{1, 2, 3\}\}$.

A *graph* G is a pair (V, E), where V is a non-empty set of *vertices* of G and E a subset of $\mathcal{P}_2(V)$, called the *edges* of G. Since $\{u, v\}$ occurs only once in $\mathcal{P}_2(V)$, and as a set $\{u, v\}$ is the same as $\{v, u\}$, according to this definition, graphs are *simple* and *undirected*. Pictorially, an edge $\{u, v\}$ is represented as in Figure 13.5.

Some familiar examples of graphs are *paths* P_n, *cycles* Z_n, *complete graphs* K_n (See Figure 13.6).

However, one of the most interesting graphs is the *Petersen graph* on ten vertices (see Figure 13.7).

A graph is a *connected graph* if there is a path between every pair of its vertices. A *subgraph* of a graph $G = (V, G)$ is a graph $H = (W, F)$ with W a non-empty set of V and a subset F of some edges $\{u, v\}$ in E with u, v in W. The set F of edges of H is allowed to be empty, in which case, it is *totally*

FIGURE 13.5: An edge of a graph.

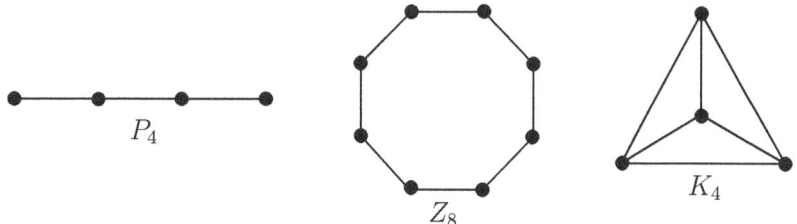

FIGURE 13.6: Paths, cycles, and complete graphs.

disconnected. On the other hand, for the *complete graph* K_n on a set V of n vertices, we take $E = \mathcal{P}_2(V)$. A maximal connected subgraph H (with as many vertices and edges as possible) of a graph G is a *connected component* of G.

Generally, it suffices to focus the attention only on connected graphs. So, from now on, graphs considered are simple, undirected, and connected. Euler had also proved the so-called *First Theorem of Graph Theory*.

• For a finite graph $G = (V, E)$, *i.e.*, *a graph on a finite number of vertices*,

$$\sum_{v \text{ in } V} \deg(v) = 2\,\mathrm{Card}(E).$$

Traveling Salesman Problem

The Traveling Salesman Problem asks if a given (connected) graph G contains a cycle Z_n, called a *Hamiltonian cycle*, going through each of n vertices of G exactly once. A graph is a *Hamiltonian graph* if it contains a Hamiltonian cycle. Clearly, every cycle is Hamiltonian. Figure 13.8 exhibits two graphs: one Eulerian, but not Hamiltonian, the other Hamiltonian, but not Eulerian.

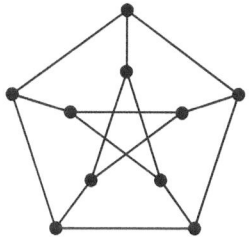

FIGURE 13.7: Petersen graph.

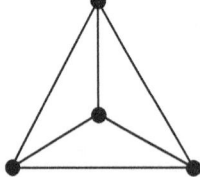 

FIGURE 13.8: Hamiltonian vs. Eulerian.

Planar Graphs

A graph is *planar* if its pictorial representation can be drawn on a plane, equivalently on a sphere, without edges crossing. Paths, cycles, K_4 are planar graphs. The simplest examples of non-planar graphs are the complete graph K_5, the *bipartite graph* $K_{3,3}$ (see Figure 13.9) and the Petersen graph.

By Kuratowski's Theorem, K_5 and $K_{3,3}$ are essentially the only non-planar graphs.

On the other hand, the skeletons of all polyhedra are planar graphs (a *polyhedron* is a three-dimensional convex shape with polygonal faces, straight edges, and sharp corners, which are the vertices of its skeleton graph). Some of the well-known polyhedra are the Platonic solids (tetrahedron, cube, octahedron, dodecahedron, and icosidodecahedron).

A *topological space* is a shape, and two topological spaces are *homeomorphic* (or essentially equal $\cong$) if one can be transformed in a continuous way to the other and vice versa without tearing them apart or poking holes in them. As the saying goes, a topologist cannot distinguish between a cup, a doughnut, and a sphere with a handle (see Figure 13.10).

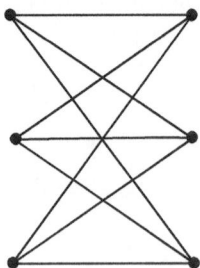

FIGURE 13.9: Bipartite graph $K_{3,3}$.

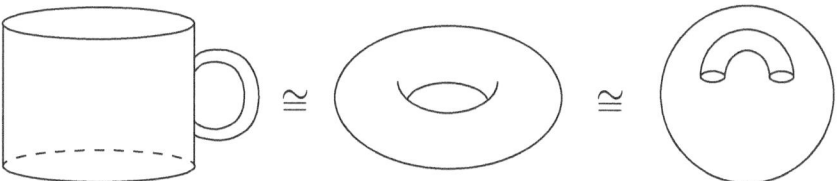

FIGURE 13.10: Homeomorphic spaces.

If n_j is the number of j-dimensional objects (i.e., points, edges, and faces for $j = 0, 1, 2$, respectively) of a Platonic solid, or that of the corresponding planar graphs, Euler discovered and proved the following amazing fact:

$$\bullet \quad n_0 - n_1 + n_2 = 2 \qquad (13.4)$$

or in Euler's own notation,

$$\boxed{V - E + F = 2.}$$

Actually, this holds for all planar graphs. Moreover, all the faces need not be bounded by the same number of edges, because removing (or inserting) an edge decreases (increases) n_1 and n_2 by 1 and thus $V - E + F$ remains unchanged. This hint should more than suffice for the proof of the following generalization of (13.4) by induction on the number of edges or vertices of a graph:

• *If n_0, n_1, and n_2 is the number of vertices, edges, and faces, respectively, of a planar graph, then*

$$n_0 - n_1 + n_2 = 2.$$

Euler-Poincaré Characteristic

Henri Poincaré (1854–1912) is another great mathematician who made indispensable contributions to topology. After Euler, he may be the most original topologist ever. One of the most studied invariant in mathematics is the so-called *Euler-Poincaré characters*, which we now explain.

A *torus* of *genus* $g \geq 0$ is a sphere with g handles (see Figure 13.11). The sphere is a torus of genus zero, doughnut is a torus of genus 1, and so on.

Suppose a graph is not planar. While drawing it on the sphere, two edges must cross each other. To avoid an edge crossing, a handle or a bridge is attached to the sphere (see Figure 13.12).

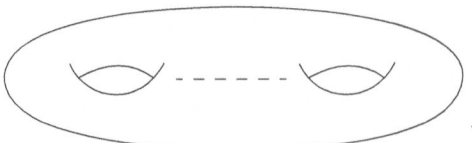

FIGURE 13.11: A torus of genus g.

Thus, any (finite) graph can be drawn without its edges crossing, on a torus of a large enough genus. The smallest $g \geq 0$ such that the graph G can be drawn without its edges crossing on a torus of genus g, is the *genus of the graph* G and is denoted by $g(G)$. It is not too hard to figure out that $g(G) = 1$ for $G = K_5$ – and the Petersen graph,

The *Euler-Poincaré characteristic* of a (connected) graph G of genus g is the integer

$$\chi(G) = n_0 - n_1 + n_2$$

whereas before, n_0, n_1, n_2 are respectively the number of vertices, edges, and faces of G.

Euler Characteristic Formula

• *If G is a connected graph of genus g, then its Euler-Poincaré characteristic*

$$\boxed{\chi(G) = 2 - 2g.}$$ (∗)

The proof is again by induction, but now on $g = g(G)$. We already have it for $g = 0$, the planar graphs. Skipping the details, the main idea in the proof is the following.

To avoid edge crossing, if necessary, a handle is attached to the previous torus with $g - 1$ handles over two triangular faces (see Figure 13.13).

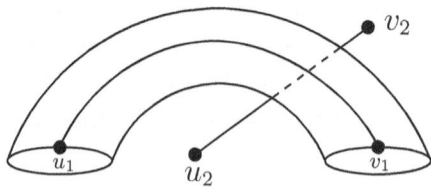

FIGURE 13.12: Bridge to avoid edge crossing.

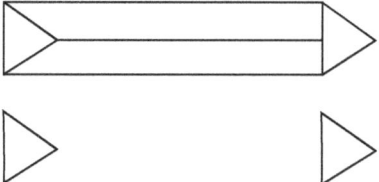

FIGURE 13.13: A handle for a torus.

This eliminates two faces but adds three. Thus, the new n_2 goes up by 1. The new n_1 goes up by 3 due to the addition of 3 edges on the handle. Therefore, $\chi(G) = n_0 - n_1 + n_2$ goes down by 2, which by induction proves (∗). □

A three-dimensional object in the above discussion would be a Platonic solid. There is a way to define an n-dimensional analog X of these objects called an *n-simplex* for any $n \geq 0$. If n_j is the number of subsimplices of X of dimension j, the Euler-Poincaré characteristic of X is the integer

$$\chi(X) = \sum_{j \geq 0} (-1)^j n_j. \tag{13.5}$$

During the 20th century, a large number of good mathematicians spent a lot of time studying $\chi(X)$ defined in (13.5). In algebraic topology, the integers n_j are replaced by the so-called *Betti numbers* β_j. It is then proved that:

• $\mathbb{R}^m \cong \mathbb{R}^n$ *if and only if $m = n$. In particular, the Euclidean plane cannot be homeomorphic to either the line or the Euclidean 3-space $\mathbb{R}^3$.*

Brower's Fixed Point Theorem has the interesting corollary:

• *If $f : B \to B$ is a continuous function from a closed ball B in $\mathbb{R}^n$ ($n \geq 1$) to itself, then it has a fixed point (x in B with $f(x) = x$).*

• *If you drop a map of, say, lower 48 anywhere therein, a point on the map will be exactly above the point it represents.*

For a quick introduction to graph theory and algebraic topology, [Mer] and [Nar] are good sources.

14

Non-European Roots of Mathematics

During the past few decades, the Eurocentric approach to the history of mathematics has undergone a considerable revision to take into account the role of other civilizations in the development of mathematics (see [Jos]). After giving a brief introduction to these civilizations, we shall delve into some mathematics having roots therein.

Arabia

The discovery of oil in the Middle East and the politics related to its control is a recent phenomenon. The region has a very rich and profound history. We have already pointed out the contribution of Babylon to mathematics. In this chapter we shall discuss, from a cultural perspective, the mathematical life in Arabia during the medieval period. For an excellent account of Islamic mathematics, see [Ber]. The mathematics of Arabia is a result of the social conditions prevalent during roughly the period AD 600–1200. Prophet Mohammed was born in 570; 610 is the year of the birth of Islam. In 630, Mohammed captured Mecca and Islam had already attracted the allegiance of the inhabitants of the Arabian Peninsula. Muslims had their capital originally in Damascus, but soon the empire split into the so-called caliphates whose rulers were called caliphs. By the year 711, Muslims had already entered Spain, conquering Egypt and the territory along the northern coast of Africa on the way. In the east, Persia had already been captured by 650, and soon the victorious Muslim armies had reached central Asia and India. At that time, India and Iran (Persia) were neighbors, meeting somewhere around the middle of the present-day Afghanistan.

In 766, Baghdad was founded by Caliph al Mansur; and soon Baghdad became a prosperous city and center of intellectual activity. In 773 an Indian scholar visited Caliph al Mansur and presented him with Indian texts on *Surya Sidhanta* (astronomy), trigonometry and arithmetic. In 800, the Library was founded in Baghdad to employ the Greek scholars who were fleeing persecution in their homelands; and thus began the translation into Arabic of Greek and also of Indian works on mathematics and astronomy. In 820 Bayt al Hikam (the **House of Wisdom**) was founded. An Institute for Advanced Studies, it lasted 200 years; and attracted scholars from all over. It was here that the works of Euclid, Archimedes, Apollonius, Diophantus and Ptolemy, as well as the Hindu texts on arithmetic and astronomy were translated into Arabic. Were it not for these translations, we would have lost forever the Greek mathematics we know today. In fact, it is believed that a good deal of Greek mathematics had already been lost when the Mouseion in Alexandria was set on fire. It is worth noting that the Muslim scholars dedicated their work to Allah, and thus secular learning was not viewed as contrary to the teaching of Islam.

Although Muslim scholars made original contributions to mathematics, their indispensable contributions have been

(i)   the preservation of Greek mathematics, and

(ii) the transmission of Hindu contributions to algebra, arithmetic and trigonometry to Europe. Probably the most durable item they passed on from the East to West is our decimal number system from Hindus, called the *Arabic* or *Arabic-Hindu numerals*, and to a lesser extent the game of chess.

The etymology of many of our mathematical terms is intimately connected with their journey from the East to West, which we illustrate with the etymology of three words: **rank, sine** and **zero.**

A plausible etymology of the word rank is this. In most Indian languages and in Persian the word *rang* means color but was used on the way to Europe to exclusively signify the color of the uniform of a soldier, and hence his *rank* in the army. In French and German the word for rank is still *rang*. It is an ironic twist of history that the word rank was brought back to India by the British Empire, apparently with no connection to the original *rang*.

The etymology of sine (Chapter 12, equation (12.1)) is more complicated. It started with the Sanskrit word *jya-ardh* (chord-half or half-chord). In the 5th century AD, Indian mathematician Aryabhata or Aryabhat (pronounced Arya

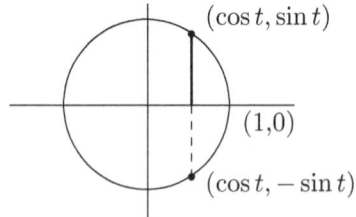

FIGURE 14.1: Aryabhat's $\sin t$ as *jya-ardh.*

Bhatt), considered to be one of the pioneers in the development of trigonometry, coined the term *jya-ardh* for sine. It is evident (see Figure 14.1) that he was using the conceptually better definition of trigonometric functions, using the unit circle, as functions of a real variable and not the one given in many trigonometry books, as the ratio of the sides of right triangles. (When asked the value of $\sin t$ for $t = 90$, not $t = 90°$, most students answered 1, which is wrong. The correct answer is the y-coordinate of the end point after traveling, counterclockwise, 90 units on the unit circle.) Aryabhat abbreviated *jya-ardh* as *jya* or *jiva*. Later, his works were translated to Arabic. Recall that Arabs pronounce v and p as b. For example, Arabs pronounce people as beable and Pepsi as bebsi. In Arabic, the term *jiva* was thus phonetically translated into otherwise meaningless *jiba* and abbreviated as jb, the same way we would abbreviate rank as rk. When later Arab writers tried to decipher the abbreviation, the closest meaningful word they could think of was *jaib*, meaning something protruding, like a pocket, nose or breast. In the 12th century when an Arabic text on trigonometry was translated into Latin, the translator used the equivalent Latin word *sinus* for nose.

Zero started as *sunya* in Sanskrit, meaning empty. It was translated to Arabic as *sifr*, and from Arabic to Latin as *cipher* (zephirum in medieval Latin).

Among all Muslim mathematicians, the most notable was probably **Mohammed al Khwarizmi** (780–850) from Khwarizm in Uzbekistan. He was among the first scholars at the House of Wisdom. He was an astronomer, geographer, and above all an eminent mathematician. The word *algebra* is derived from the title of his book *al Gebr W'al Muqabala*. *Al gebr* roughly means "restoring" and *muqabala* means "comparison." Put together, the title means "transferring terms from one side of the equation to the other." During the 13th century, this book was translated into Latin. So was *Liber Algorismus* (al Khwarizmi's book) the origin of another word: *algorithm*. The earliest available Arab text on arithmetic dealing with Hindu numerals is also al Khwarizmi's *Kitab al-jam W'al tafriq bi Hisab al Hind* (Book on the Arithmetic of Addition and Subtraction as Done in India).

Perhaps one of the most prolific Arab mathematicians, **Ibn al Haytham** (965–1039) was known in Europe as "Alhazen." He was born in Basra but spent most of his life in Egypt, where he wrote seven books on *Optics* that were translated into Latin during the 13th century. He also wrote other classics. For details, see [Hog-2].

Al-Biruni (973–1055), also from Uzbekistan, wrote on astronomy, geography and mathematics. During the latter half of his life he traveled to India and wrote extensively on Indian culture, Hindu philosophy and the rules of chess.

Two famous Persian and Muslim mathematicians are al Karaji and **Omar Khayyam** (1048–1122). Khayyam worked on cubic equations, but he is best known for the collection of poems titled *Rubaiyat*. In Chapter 10, we mentioned al Karaji for initiating the congruent number problem.

Decimal Fractions

Around AD 950, Abu'l-Hasan al-Uqlidisi, who lived in Damascus, wrote *The Book of Indian Arithmetic* on decimal fractions. Little is known about the author except that he made a living copying works of Euclid (hence the name Uqlidisi). The idea of decimal fractions was further developed by **al Samawal** (1125–1174) and completed by al Kashi (1380–1429). It is believed that a Turkish colleague of al Kashi, known as Ali Qushji, transported this knowledge to Turkey. In 1562, this knowledge appeared in central Europe as a collection of Byzantine problems. It is quite likely that Simon Stevin (1548–1620), who is often credited in the West with inventing decimal fractions, had learned the Turkish method from this Byzantine text. For details, see [Jos, pp. 317–318].

China

Although China was not always a politically united country, culturally, we can speak of it as one nation throughout its history. The Chinese society was governed not by religion in the western sense, but by a code of ethical conduct, called *Confucianism*, after the philosopher Confucius who lived around 500 BC *Neo-Confucianism* combined these moral standards with elements of Buddhism and Taoism.

By the time of the founding of the Shang dynasty around 1750 BC, a considerable civilization was already in place. It had arisen in the valley of the Yellow River, and from there it spread south, east and west. The artifacts found in the graves of Shang rulers in the Hunan province are from this civilization. The silk, metal and pottery industries were already well established. The Shangs had a kind of script that has been found inscribed on bones and sea shells.

About 1050 BC, the Zhou people of western China replaced the Shang dynasty with one of their own which lasted until 250 BC This was followed by the short-lived (about 20 years) Qin dynasty that brought great changes to the social order. Weights, measures, currency and the Chinese system of writing were standardized and the Great Wall of China was built.

During the Han period (200 BC–AD 220), paper was invented in China and a civil service, based on imperial examinations rather than family ties, was instituted. The system of imperial examinations continues, with a few short periods of disruption, to this day. During a period (AD 220–600) of disunity, non-Chinese nomads from the north invaded and ruled China. In the Han period, Buddhism spread across China and into all aspects of life.

The Tang dynasty, founded in AD 618, lasted 300 years. It was followed by the Sung dynasty (960–1279). Mongol warrior Genghis Khan's grandson, Kublai Khan, founded the Yuan dynasty in 1279. It was Kublai Khan whose court was visited by Marco Polo. Kublai Khan died in 1294.

The Ming dynasty ruled from 1368 to 1644, a period of stability, prosperity and Chinese influence in eastern Asia. In 1644 Manchus from Manchuria invaded China and established the Qing dynasty which ruled China until 1912.

Unlike Brahmins in India who formed the most privileged class and were free to pursue whatever scholarly activity they liked, Chinese mathematicians had to pass examinations and were employed by the rulers. The demands of the empire included surveying, calendar making and computation of taxes. The imperial government thus needed the study of applicable mathematics. In fact, at various times there was an Imperial Institute of Mathematics whose officials were trained in practical mathematics.

The multiplication table for numbers up to nine was known in China from time immemorial. During the Han dynasty, Chinese knew the Pythagorean theorem and triplets, as well as the formula for the area of the circle discussed in Chapter 6. They used base ten and could extract square and cube roots. They approximated the value of π as $\frac{22}{7}$ and $\frac{355}{113}$. *The Ten Books of Arithmetic* came into being between the 2nd century BC and the AD 6th century, edited by Liu Hui and others. The most important of these is *Chiuchang-Suanshu* which contains the *Nine Chapters*. **Liu Hui** (AD 3rd century) was an expert on the mathematics of surveying. These books were used in civil service examinations up to the Sung era. They contained a treatment of positive and negative fractions and a number of problems reducible to equations of 3rd and 4th degree. **Sun Zi** (late AD 3rd century) studied the Chinese remainder problem that we shall discuss in the next section. **Qin Jiushao** (1202–1261) studied roots of polynomials and linear congruences in up to five variables, whereas **Li Xi** (1192–1279) applied algebra to geometry. **Matteo Ricci** (1552–1610) introduced European mathematics to China, and traditional Chinese mathematics declined. For more information, see [Mar].

Chinese Remainder Theorem

In number theory, the most important problem Chinese mathematicians studied is called the Chinese remainder problem. Its solution has widespread application in pure mathematics. It arose as follows. From time immemorial there has been a full moon every 28 days. It is for this reason that the Hebrews called a period of 28 days a *lunar month*. A quarter of a lunar month consisting of 7 days is a *week*. Imagine a metric enthusiast insisting on a week of 10 days. Suppose an important event occurred on, say, the 4th day of the lunar week and on the 8th day of the metric week. Is it possible to calculate the day x on which the event took place? After a lapse of $70 = 7 \cdot 10$ days, both weeks start all over again with day 1. So if x_0 is an answer with $0 \leq x_0 < 70$, so is $x = x_0 + 70k$ for all k in $\mathbb{Z}$. In the notation invented by C.F. Gauss, which we discussed in Chapter 4, the Chinese remainder problem can be stated as follows: For what integer x $(0 \leq x < 70)$, is it true that

$$\left. \begin{aligned} x &\equiv 4 \pmod 7 \\[1mm] \text{as well as} \qquad \\[1mm] x &\equiv 8 \pmod{10}? \end{aligned} \right\} \tag{14.1}$$

Recall an integer x *is congruent to y modulo m* if $x - y$ is a multiple of m. Gauss wrote this relationship as

$$x \equiv y \pmod m.$$

For example, $18 \equiv 4 \pmod 7$ and $18 \equiv 8 \pmod{10}$. If $x, y > 0$, this amounts to saying that $x \equiv y \pmod m$ if and only if x and y have the same remainder on division by m. If $0 \leq a < m$, then "a is the remainder of x on division by m" can be abbreviated as

$$x \equiv a \pmod m.$$

If $m = 7$, then the remainder a represents the day of the week.

Recall also that for a fixed $m > 0$, called the *modulus*, the relation $x \equiv y$ $\pmod m$ partitions the set $\mathbb{Z}$ of integers into m disjoint subsets, called the *residue classes* mod m. These residue classes mod m are the sets $\bar{r} = r + m\mathbb{Z}$ with $0 \leq r < m$. We now state the **Chinese Remainder Theorem** (CRT).

- *Suppose we are given r integers $m_1, \ldots, m_r > 1$, coprime in pairs (no two of them have a common factor larger than 1) and integers $a_1, \ldots, a_r$ $(0 \leq a_j < m_j)$. Let $m = m_1 \ldots m_r$. There is a unique x $(0 \leq x < m)$ such that for every j,*

$$x \equiv a_j \pmod{m_j}.$$

(Note that in our earlier example, the integers $m_1 = 7$ and $m_2 = 10$ had no common factor, so that the hypothesis of the Chinese Remainder Theorem was satisfied.) If x is a solution, then so is $x + km$ for all k in $\mathbb{Z}$. To compute the solution x, we need the following consequence of the Euclidean Algorithm:

• *Suppose b and m are positive integers $(m > 1)$ with no common factor and $0 \le a < m$. Then there is an integer y $(0 \le y < m)$ such that*

$$by \equiv a \pmod{m}.$$

Proof. By the Euclidean Algorithm, we can write the GCD. $(b, m) = 1 = by + mx$ for some x, y in $\mathbb{Z}$. Hence $by \equiv 1 \pmod{m}$. Ignoring multiples of m in y, we can assume $0 \le y < m$.

The number y is found, in general, by trying all numbers from zero to $m - 1$, if m is small. For large m, of course, the Euclidean Algorithm is indispensable. □

Examples.

1. Let $b = 10$, $a = 4$ and $m = 7$. Then $y = x_1 = 6$ is a solution of

$$10x_1 \equiv 4 \pmod{7}.$$

2. If $b = 7, a = 8$ and $m = 10$, then $y = x_2 = 4$ is a solution of

$$7x_2 \equiv 8 \pmod{10}.$$

If we put $x = 10x_1 + 7x_2$, then the remainder of x on division by 10 comes from $7x_2$ which is 4 and the remainder of x on division by 7 is the same as that of $10x_1$ which is 8. Thus the solution of (14.1) is

$$x = 10 \cdot 6 + 7 \cdot 4 = 88.$$

The answer in the range $0 \le x < 70$ is of course $88 - 70 = 18$.

This can be generalized not only to prove the Chinese Remainder Theorem, but also to find the solution x.

Proof of the Chinese Remainder Theorem.. The hypothesis implies that for every j, m_j and $\frac{m}{m_j}$ have no common factor. Hence by the Euclidean Algorithm, there is an x_j $(0 \le x_j < m_j)$ such that

$$\frac{m}{m_j} \cdot x_j \equiv a_j \pmod{m_j}.$$

Clearly $\frac{m}{m_i} x_i$ is a multiple of m_j for every $i \ne j$. Hence its remainder on division by m_j is zero. So if we put

$$x = \frac{m}{m_1} x_1 + \cdots + \frac{m}{m_r} x_r,$$

we have a simultaneous solution of all the congruences

$$x \equiv a_j \pmod{m_j}.$$

The uniqueness is obvious, because for both x, y to be solutions ($0 \le y \le x < m$), $x - y$ has to be a multiple of m. This can happen only if $x = y$.

Example. We work out the solution of (14.1) again to illustrate this process. We first solve

$$\frac{7 \cdot 10}{7} x_1 \equiv 4 \pmod 7 \text{ and } \frac{7 \cdot 10}{10} x_2 \equiv 8 \pmod{10}$$

that is

$$10x_1 \equiv 4 \pmod 7 \text{ and } 7x_2 \equiv 8 \pmod{10}.$$

We have already found above that $x_1 = 6$ and $x_2 = 4$. Hence

$$x = \frac{7 \cdot 10}{7} \cdot 6 + \frac{7 \cdot 10}{10} \cdot 4 = 88 \equiv 18 \pmod{70}. \qquad \square$$

Exercises.

Solve the following simultaneous congruences:

$x \equiv 2 \pmod 3$	2. $x \equiv 4 \pmod 5$
$\equiv 3 \pmod 5$	$\equiv 6 \pmod 8$
$\equiv 4 \pmod 7$	$\equiv 8 \pmod 9$

(The second problem is called the problem of Master Sun, after Sun Zi.)

Application*

Recall Euler's *totient function* $\phi(m)$. This function plays a central role in algebra and number theory. For an integer $m \ge 1$, $\phi(m)$ is the number of integers a ($0 < a \le m$) such that a and m have no common factor larger than 1. Table 14.1 lists $\phi(m)$ for a few values of m.

We prepared this table by just counting the numbers of a ($0 < a \le m$) such that a and m have no common factor. Is there an easier way to compute $\phi(m)$? If $m = p^e$ is a prime power, then there is one. For a to have a factor in common with m, it has to be of the form pb ($1 \le b \le p^{e-1}$). Hence $\phi(p^e) = p^e - p^{e-1}$ or

$$\phi(p^e) = p^e \left(1 - \frac{1}{p}\right).$$

TABLE 14.1: Some values of $\phi(m)$.

m	$\phi(m)$
1	1
2	1
3	2
4	2
5	4
6	2
7	6
11	10
12	4

It is a consequence of the CRT that the Euler ϕ-function is multiplicative, i.e., if GCD $(m,n) = 1$, then $\phi(mn) = \phi(m)\phi(n)$. Thus if we write

$$m = p_1^{e_1} \ldots p_r^{e_r}$$

as product of powers of distinct primes $p_1, \ldots, p_r$, then

$$\phi(m) = \prod_{j=1}^{r} \phi(p_j^{e_j})$$

$$= \prod_{j=1}^{r} p_j^{e_j} \left(1 - \frac{1}{p_j}\right)$$

$$= m \prod_{j=1}^{r} \left(1 - \frac{1}{p_j}\right).$$

And we obtain the well-known result on Euler's ϕ-function in number theory.

$$\bullet \qquad \phi(m) = m \prod_{p|m} \left(1 - \frac{1}{p}\right).$$

In the above formula for $\phi(m)$, the product is over all prime divisors p of m. For example,

$$\phi(12) = 12 \left(1 - \frac{1}{2}\right)\left(1 - \frac{1}{3}\right) = 4.$$

This coincides with the value of $\phi(12)$ in Table 14.1.

Example. If $m = 60 = 2^2 \cdot 3 \cdot 5$, its prime divisors are $p = 2, 3$ and 5. Hence $\phi(60) = 60 \cdot \left(1 - \frac{1}{2}\right)\left(1 - \frac{1}{3}\right)\left(1 - \frac{1}{5}\right) = 16.$

Exercise. Show that $\phi(m)$ given by its definition and by the formula in the theorem above, coincide for all m in the range $13 \leq m \leq 40$.

For more on Chinese mathematics, see [Mar].

India

India and China have similar histories in more than one way. They have the oldest uninterrupted civilizations in the world. Both assimilated new ideas from wherever they came. In the West (that is west of the Khyber Pass in the Hindu Kush Mountains), new religions tried to wipe out their predecessors. Of course, there were wars both in China and India, but until the arrival of Muslims from the West, the wars were over territory and not religion. To understand the origin of Indian mathematics, also called the Hindu mathematics, a rudimentary knowledge of Indian history is essential, which we divide into four periods. (It must be mentioned that until recently, Indians did not take history seriously. In fact, the early history of India itself has been recorded only by Chinese travelers. Thus some historical claims about Indian science may be open to debate.)

1. *Prehistoric.* As has already been mentioned in Chapter 1, this period is roughly 3000 BC–1500 BC. The civilizations of Harappa and Mohinjodaro along the Indus River belong to this period. Recent excavations have found the existence of well-planned cities, which suggests that these people possessed a functional knowledge of arithmetic and geometry.

2. During the period 1500 BC–1000 BC, waves of immigrants from central Asia settled in north India. Some of these tribes from central Asia also went west to Europe. The descendants of these immigrants both in India and in Europe are referred to as *Indo-European* or *Indo-Germanic*. In India they developed a distinctly Indian culture called *Hinduism*.

In the western sense of the word, Hinduism is not a religion but a continuously evolving way of life. Unlike a typical western religion, Hinduism is not the creation of one man or prophet. It is a collective philosophy. This explains the attitudes of Hindus toward other religions. Two persons may do totally different things or worship different gods but still both would consider each other good Hindus. The concept of conversion is totally alien to Hindus and Buddhists alike. However, they welcome the exchange of ideas. There is no word in India for religion in the Western sense, the nearest one is *dharma* (duty) or the path of righteousness. In Hinduism what one says or believes in is of no consequence. It is *karma* (deeds) that counts.

Early Hindus organized their society into four groups based exclusively on profession. *Brahmins* were the scholars and *Kshatryas* (also known as Khatris in Punjab) were the soldiers, while *Vashyas* (farmers and traders) formed the backbone of economy. Finally, the natives who were conquered, like everywhere else, were to serve Brahmins, Kshatryas and Vashyas. These unfortunate people were put into the lowest group, and were called *Sudras*.

In the beginning except for Sudras, changing professions was possible and frequent. But with the passage of time, as there were no formal institutions for job training, the tricks of the trade were passed on to the next generation by the parents, and it became increasingly difficult to go into a profession other than that of one's parents. Later on, the rigidity of this division came to be known as the *caste system*. Each caste developed its own subculture, and intermarriages between different castes ceased to exist.

3. *Golden Period.* The language of Brahmins was Sanskrit, which is also the mother of most modern Indian languages. As has already been said, the roots of many mathematical terms go back to Sanskrit. The most productive period of Hinduism was roughly the *golden period* 1000 BC–AD 1000. At the beginning of this period great epics like *Vedas* were written, and the foundation of Hinduism was firmly laid down. A little later, Buddha (563–483 BC) appeared. During the first half of the first millennium AD, our decimal number system was invented. Ashoka the Great ruled India (247–236 BC) and spread Buddhism to China and its neighboring countries. This inevitably resulted in the exchange of ideas, including mathematics, between India and China.

In 327 BC, Alexander the Great invaded India. He brought with him not only soldiers, but learned men also. This ushered in an exchange of ideas between Greeks and Indians. The works of Hipparchus had an impact on Indian astronomy. Two of the oldest universities in the world, Nalanda (in Bengal) and Taxila (in the Punjab), belonged to this period. Apollonius was one of the distinguished visitors to Taxila. The Greek influence can also be seen on Indian sculptures. Emperor Ashoka set up inscribed pillars at important places in his kingdom. The most famous was the one from Sarnath where Buddha preached his first sermon.

Our number system was invented during the first half of the first millennium. The symbol 0 was invented a little later–first, to indicate an empty place in our place-value number system and then to denote the number zero. The first known use of zero as a number was in the Bakhshali manuscript of AD 3rd century. As a number, zero was defined by

$$a \pm 0 = a, a \cdot 0 = 0 \text{ and } \frac{0}{a} = 0.$$

Division by zero was not permitted; but in symbolic arithmetic, $\frac{a}{0}$ $(a > 0)$ was made to play the role of ∞.

4. *Period of Muslim invasions.* Until about AD 1000, all Indians (Hindus, Buddhists, Jains and others) not only lived in peace and harmony, but complemented each other's spiritual needs. This harmony was shattered by the Muslim incursions from the northwest. Forcible conversions caused great suffering and slowed down, then halted, the scholarly activity in India. On the other hand, Arabic culture and literature influenced the local tradition, making India culturally very diverse and rich. In the year 1026 the first Muslim invader, Mehmood Ghazni, destroyed Hindu temples and looted the gold therein. Others followed, including, in 1520, Babar (a descendant of Tamerlane from central Asia), who founded the Mughal Empire, which lasted 200 years. It is during the Mughal period that great architectural wonders like the Taj Mahal were built.

5. Finally, the British replaced the Mughals as the absolute rulers of India. Their conquest of India began with the establishment of the East India Company in Calcutta (1600). It is perhaps to the credit of the British democracy that the British showed a greater respect for Indian customs and religions than did Muslim invaders. The British left in 1947. During their rule, they introduced their system of education and bureaucracy. The former provides India with one of the most skilled workforces in the world, while the latter is a formidable hindrance to its development. During the 20th century, **S. Ramanujan** (1887–1920), a self-taught mathematician, had no rival in India; some would say anywhere in the world.

Indian colleges and universities were patterned after those in England. There was not much emphasis on conceptual mathematics. The mathematics taught was mostly utilitarian, such as that needed for Newtonian physics. After independence in 1947, the leading mathematicians in India started to look to America, France and Germany for setting up new institutions of higher learning and research. In particular, the Tata Institute of Fundamental Research was founded in Bombay as a national center for mathematics. It was heavily influenced by the French school of algebraic geometry. Lately the IITs (Indian Institutes of Technology) have earned a good reputation in informatics, and to a lesser extent, in mathematics. Lately, the Indian Institutes of Science Education and Research (IISERs) have been added to this list.

Golden Period

It may be because of the humility of the Hindus that we do not know the names of the scholars who invented our decimal number system. Prior to the modern era, it is the classical period (roughly the millennium AD 500–1500), considered by some as the golden period of Indian mathematics, from which we can name some prominent mathematicians from India. The first identifiable Hindu mathematician was **Aryabhat** (AD 5th century) from Patna in the eastern part of India. (Aryabhat is usually spelled as Aryabhata, but the actual pronunciation is closer to bhat than to bhata.) He wrote on astronomy and trigonometry. The origin of the word *sine* has been traced back to him. He also calculated the value of π as 3.1415, as well as studied the roots of a quadratic equation.

In Chapter 3, we showed how Euclid computed the greatest common divisor $d = (a, b)$ of a and b, which was abbreviated as GCD (a, b). However, Euclid did not consider the representation $d = \lambda a + \mu b$ as a linear combination of a and b. The first mathematician to use the Euclidean Algorithm to solve the Diophantine equation

$$ax + by = d,$$

was Aryabhat (see [Wei-1, p. 7]). It is obvious that for this Diophantine equation to have a solution, it is necessary and sufficient that d be a multiple of the GCD (a, b). Of particular interest is the Diophantine equation

$$ax + by = 1,$$

where GCD $(a, b) = 1$. It plays an important role in algebra and number theory. For example, a solution of the Chinese remainder problem is obtained from that of $ax + by = 1$. Another application is the computation of multiplicative inverses of non-zero elements of finite fields, as discussed in Chapter 4.

Aryabhat was followed by **Brahmagupta** (born in AD 598) from Rajasthan. He also wrote on astronomy but is best known for his work in algebra. His quadratic formula is virtually as we know it today, that is, *the two solutions of the quadratic equation*

$$ax^2 + bx + c = 0 \ (a \neq 0)$$

are

$$x = \frac{-b \pm \sqrt{b^2 - 4ac}}{2a}.$$

(We are not saying the solutions were not known elsewhere, say, in Babylon or in Greece.) We have already mentioned his generalization of Heron's formula for the area A of a triangle to that of the circular quadrilateral:

$$A = \sqrt{(s - a)(s - b)(s - c)(s - d)},$$

where s is the semi-perimeter $s = \frac{a+b+c+d}{2}$. A *circular quadrilateral* is a quadrilateral inscribed in a circle.

Pell's Equation

As mentioned earlier, Archimedes, in his letter to Eratosthenes, was probably the first mathematician to come up with a Pell equation; but Brahmagupta was the first mathematician to study it systematically, almost as we would do it today. (See the fascinating book by André Weil [Wei-1]). Let $m > 1$ be a positive integer with no non-trivial square factor, and let $s = (x, y)$ and $s' = (x', y')$ be two given solutions of the Pell equation, i.e.,

$$x^2 - my^2 = 1. \tag{14.2}$$

Brahmagupta defines *bhavana* or a binary operation on the set of solutions of (14.2) by

$$s \cdot s' = (x, y) \cdot (x', y') = (xx' + myy', xy' + x'y). \tag{14.3}$$

Exercise. Prove that the RHS of (14.3) is also a solution of (14.2).

Notice that $e = (1, 0)$ is always a solution, which we call the *trivial solution*. If we have a non-trivial integer solution (x_1, y_1), that is, one with $x_1 > 0, y_1 > 0$, Brahmagupta argued that by his bhavana, we can produce more and more solutions, namely:

$$(x_2, y_2) = (x_1, y_1)^2 = (x_1, y_1) \cdot (x_1, y_1)$$
$$= (x_1^2 + my_1^2, 2x_1y_1),$$

$$(x_3, y_3) = (x_1, y_1)^3 = (x_1, y_1) \cdot (x_2, y_2)$$
$$= (x_1x_2 + my_1y_2, x_1y_2 + x_2y_1)$$

and so on. Since m, x_1, y_1 are all integers ≥ 1, it is obvious that the integers $y_1, y_2, y_3, \dots$ satisfy

$$y_1 < y_2 < y_3 < \cdots .$$

Brahmagupta was thus able to claim that one can obtain infinitely many integer solutions starting from a non-trivial one. Brahmagupta's work on equation (14.2) can be stated as follows.

- *If equation* (14.2) *has one non-trivial integer solution, it has infinitely many.*

One can translate the work of Brahmagupta into the modern language (of group theory).

So let us return to the equation $x^2 - my^2 = 1$, where m is a positive integer. We may restrict x to be positive, that is, consider only the (integer) points on the hyperbola with $x > 0$ (see Figure 14.2). These solutions form a group under Brahmagupta's rule. Let us call this a connected component G. Brahmagupta proved that if G has one element other than $e = (1, 0)$, then G is infinite.

Exercise. If you know what a group is, show that G is a group with $e = (1, 0)$ as its identity and $(x, y)^{-1} = (x, -y)$.

If $m = c^2 n$, we may replace cy by y. Hence it is enough to consider m with no square factor bigger than 1. If G has elements other than $e = (1, 0)$, we denote the one nearest to e in the first quadrant by (x_1, y_1). Brahmagupta assumed the existence of (x_1, y_1) from which he deduced that $x^2 - my^2 = 1$ has infinitely many (integer) solutions. Five centuries later, Bhaskar (also called Baskar II) and his contemporary, another Indian mathematician, Jeyadev, independently found (x_1, y_1) for many values of m, such as $m = 13, 61$ and 67.

These results appear in Bhaskar's book titled *Siddhanta-Shiromani*. A chapter titled *Lilavati* (dedicated to and named after his daughter) is on arithmetic and contains a complete solution of the Diophantine equation $x^2 - my^2 = 1$ for many values of m. See [Var]. (Recall that Bhaskar also gave a simple proof of the Pythagorean theorem.)

The Indian school of mathematicians had an algorithm called *Chakrawala* (the circular method), to find (x_1, y_1). See [Wei-1, pp. 19–24]. The existence of (x_1, y_1) for all square-free $m > 1$ was demonstrated by the famous French mathematician Lagrange. Finally, it was **Peter G. L. Dirichlet** (1805–1859) who proved that up to sign, every solution of the Pell equation is of the form

$$(x_n, y_n) = (x_1, y_1)^n = \underbrace{(x_1, y_1) \cdots (x_1, y_1)}_{n\text{-times}} \text{ for some } n. \text{ (See [Cha-1].) Thus the}$$

Indians had completely solved the Diophantine equation $x^2 - my^2 = 1$ for the values of m they studied. As Weil noted [Wei-1, p. 24], "to have developed *Chakrawala* and to have applied it successfully to such difficult numerical cases as $m = 61$ or $m = 67$ had been no mean achievement." For further information on Indian mathematics, the reader may consult the classic [Dat].

A Fun Problem with the Pell Equation

Among all the Diophantine equations we can write down, why has the Pell, or more appropriately, the A-B (which may stand for Aryabhat-Bhaskar) equation been studied so extensively? In this section we shall try to answer this question, citing some connections the equation has to different disciplines of mathematics. Throughout this section, we assume $m > 1$ is a fixed integer with no non-trivial square factor.

(1) *Extraction of square roots.* The problem of *extracting the square root*, numerical as well as geometric, has been important since antiquity for mathematicians all over the world. Because $\sqrt{m}$ is irrational, what we mean by

extracting square roots of m is to find better and better rational approximations to $\sqrt{m}$. If one can find infinitely many integer solutions $x, y > 0$ to the
Pell equation

$$x^2 - my^2 = 1,$$

which we can rewrite as

$$\left(\frac{x}{y}\right)^2 - m = \frac{1}{y^2},$$

then for very large integers y, $\frac{1}{y^2}$ is negligible and $\frac{x}{y}$ gives a good rational
approximation to $\sqrt{m}$.

To obtain such an approximation, we first note that the Pell equation represents a hyperbola in the plane. To each solution, other than $e = (1,0)$,
correspond four solutions, one in each quadrant, as shown in Figure 14.2.
Once we have one solution $s = (x_1, y_1) \neq e$ in the first quadrant, we can
obtain infinitely many of them as

$$(x_n, y_n) = s^n, \ n = 1, 2, 3, \ldots.$$

Since x_1, y_1 and m are all positive integers, it is clear from the group law
that the points (x_n, y_n) go to infinity very rapidly. We take (x_1, y_1) to be the
smallest solution in the first quadrant in the sense it is closest to $e = (1, 0)$. It
is a non-trivial fact that then, up to sign, every solution of the Pell equation
(14.2) is in the list $s^n = (x_n, y_n)$. This, the so-called *fundamental generator*
(x_1, y_1), can be found as follows. In $1 + my^2$ put $y = 1, 2, 3, \ldots$ until it becomes
a square x_1^2 $(x_1 > 0)$. It is guaranteed to happen, eventually, by Dirichlet's
Unit Theorem. The first y for which this happens is y_1.

Example. We illustrate this for $m = 2$, that is, for the Pell equation:

$$x^2 - 2y^2 = 1.$$

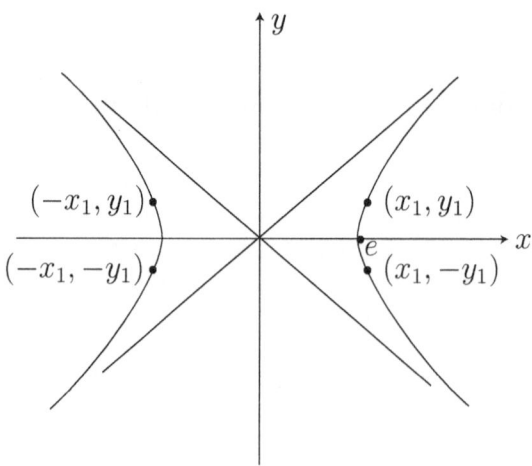

FIGURE 14.2: Hyperbola defined by the Pell equation.

For $y = 1, 1+2y^2 = 1+2 \cdot 1^2 = 3$ is not a square, but for $y = 2, 1+2 \cdot y^2 = 9 = 3^2$. Hence the generator $(x_1, y_1) = (3, 2)$. The next solution (x_2, y_2)

$$
\begin{aligned}
&= (x_1, y_1)^2 = (x_1, y_1) \cdot (x_1, y_1) \\
&= (3, 2) \cdot (3, 2) = (3 \cdot 3 + 2 \cdot 2 \cdot 2, 3 \cdot 2 + 2 \cdot 3) \\
&= (17, 12),
\end{aligned}
$$

and (x_3, y_3)

$$
\begin{aligned}
&= (x_1, y_1) \cdot (x_2, y_2) \\
&= (3, 2) \cdot (17, 12) \\
&= (3 \cdot 17 + 2 \cdot 2 \cdot 12, 3 \cdot 12 + 2 \cdot 17) \\
&= (99, 70).
\end{aligned}
$$

It is a consequence of the associativity of the group law, which we leave as an exercise that we can compute (x_5, y_5) in several ways, such as (x_5, y_5)

$$
\begin{aligned}
&= (x_2, y_2) \cdot (x_3, y_3) \\
&= (17, 12) \cdot (99, 70) \\
&= (17 \cdot 99 + 2 \cdot 12 \cdot 70, 17 \cdot 70 + 99 \cdot 12) \\
&= (3363, 2378).
\end{aligned}
$$

It is easy to check that we get the same answer if we compute $(x_5, y_5) = (x_1, y_1) \cdot (x_2, y_2)^2$.

Notes.

(1) The value $3363/2378 = 1.4142136$ of $\sqrt{2}$ is the same as the one given by a handheld calculator.

(2) It is a timely remark to point out that the Diophantine equations for other conic sections, namely

$$
ax^2 + by^2 = c \text{ or } y = ax^2
$$

representing ellipses (including circles) or parabolas are not very interesting. The first one has only finitely many integer solutions, whereas the solutions of $y = ax^2$ are parameterized by taking x in $\mathbb{Z}$. Thus from the Diophantine point of view the Pell-like equations representing hyperbolas are the only interesting ones among conic sections.

(3) Both Bhaskar and Jayadev gave a more efficient method, namely the *Chakrawala* to find the generator (x_1, y_1), see [Var, pp. 25–38].

Exercises.

1. Find the generator of the Pell equation for $m = 13, 61$ and 67 (some of the values of m dealt with by Bhaskar).

2. Using the Pell equation, compute $\sqrt{5}$, $\sqrt{6}$, and $\sqrt{7}$.

(4) *Hilbert's tenth problem* asks for an algorithm to determine, in a finite number of steps, the solvability of a given Diophantine equation. In 1970, Yuri Matiyasevich proved that such an algorithm does not exist. The proof involves the solutions of the Pell equation $x^2 + (a^2 - 1)y^2 = 1$ $(a > 0)$. (See [Bro, pp. 323–378].)

(5) *Norm form equations.* By Pell's equation one usually means a more general equation

$$x^2 - my^2 = a \qquad (14.4)$$

studied by Brahmagupta, Bhaskar and others. However, for the simplicity of exposition we have taken the *additive* $a = 1$. The equation (14.4) above can also be written as $N(\alpha) = a$, where $\alpha = x + y\sqrt{m}$ and $N(\alpha) = (x + y\sqrt{m})(x - y\sqrt{m})$ is the *norm* of α. Therefore equation (14.4) has a natural generalization to the so-called *norm form equations*. For details see [Sch-1].

Some Other Feats of Indian Scholars

It is now more or less universally accepted that our so-called Hindu-Arabic number system is a gift from India. However, apart from the game of chess, some other equally impressive feats of Indian scholars are either rarely talked about or their ingenuity not appreciated even in India, most likely due to the inferiority complex instilled in the Indian psyche by centuries of colonial rule by the British.

Not only did humanity struggle with how to record a count but also with how to record the spoken language. Whereas for recording numbers, various symbols were employed; to record words, alphabets were invented. Again, like Egyptian numerals, the Phoenician alphabet lacks something. Here also Brahmins had better ideas.

Indian scholars (Brahmins) were quite into classification and systemization of every aspect of scholarship. Panini (520–460 BC) was the first grammarian to invent grammar to describe the logical structure of a language so that it could be taught systematically.

If we look at Devanagari (script of gods), the script in which Sanskrit and some Indian languages are written, as well as the scripts derived from it for various Indian languages (Bangla, Gujarati, Punjabi, etc.), it is not arranged ad hoc like scripts derived from the Phoenician alphabet such as Arabic alph, bé, …; Greek alpha, beta; Roman A, B, …, etc. with consonants and vowels arranged haphazardly.

Devanagari is arranged in a matrix $[a_{ij}]$, where the row i in which the letter a_{ij} appears indicates the position (guttural to labial) of our vocal chord that produces the sound that corresponds to a_{ij}. There are five different tones that can be produced from each position of the vocal chord. The column j of a_{ij} indicates the pitch of the tone.

There are almost four dozen letters in Devanagari, representing vowels and consonants. The twenty-five basic consonants are arranged in a 5×5 matrix (see Figure 14.3). The shape of the letter a_{ij} is not that important, as it varies in the alphabets derived from Devanagari, e.g., for Bangla, Gujarati, and Punjabi.

It is plausible that the Russian chemist Dmitri Mendeleev (1834–1907) was inspired by Devanagari for his periodic table in chemistry. It is rather strange that the educated Indians prefer to use Roman letters over Devanagari, a phenomenon absent in China, Japan, and Korea.

As noted earlier, via Fibonacci, Arabs passed on the Hindu-Arabic number system to Europe from where it spread to all corners of the world. Perhaps because the script of the Quran is in Arabic, Arabs did not adopt Devanagari, a better script perfected by Indians. Otherwise, Fibonacci would have written another book, namely on Devanagari script beginning with, "Like the ten digits of the Indians for numbers, with four dozen letters and a dozen accents of the Indians, every sound produced by the human chord can be written and read precisely as it is, which I shall explain now." Had it happened, there would be no Spelling Bees in America. Look at how "schedule" and "lieutenant" are pronounced in America and Britain!

On the other hand, there are no Spelling Bee contests for Indian languages that use Devanagari or scripts derived from it.

FIGURE 14.3: Devanagary: the script for Sanskrit.

Part IV

Mathematics of the 20th Century

15

Mathematics of the 20th Century

The 20th century produced more mathematics than all the previous centuries combined. As for gadgets in electronics, it is impossible even to name every theorem, big or small, that has been proved during these years. The choice of topics for this chapter is not meant to slight any area of mathematics, but rather it is limited by and reflects the author's personal taste and field of expertise. This chapter is a must for graduate students aiming to specialize in number theory. Sections and paragraphs with no mathematical symbols are accessible to every reader.

Hilbert's 23 Problems

We begin with 23 problems posed by David Hilbert at the International Congress of Mathematicians at Paris in 1900, the last year of the 19th century. In essence he laid out a program for research for the next century. To discuss all these 23 problems is beyond the scope of this chapter. For a survey, see [Bro]. The very first problem Hilbert listed is the Continuum Hypothesis, which we have discussed in some detail in Chapter 5. The most famous of all of his problems is the eighth, the Riemann Hypothesis, which we discuss in the next section. The tenth problem asked for an algorithm to decide, in a finite number of steps whether a Diophantine equation has a rational solution. It was settled by Yuri Matiyasevich in 1970. He proved that there is no general method for solving every given Diophantine equation. Among all these problems, Hilbert himself considered the eighth to be the most important of all. He is quoted as saying that if he ever rises from his grave, the first thing he would want to know is if the Riemann hypothesis has been proved.

Millennial Problems

During the final year of the last millennium, to be precise on May 24, 2000, the Clay Mathematics Institute announced a \$1 million dollar prize for each of the seven millennial problems. These are considered by many distinguished mathematicians to be the most outstanding unsolved problems in mathematics. One of these is the Riemann hypothesis, which appeared a century ago in 1900 on Hilbert's list as well. In the next few sections, we will discuss it and two other millennial problems, namely, the Birch & Swinnerton-Dyer conjecture and the Poincaré conjecture. The Poincaré conjecture was proved by Grigori Perelman in 2002. He was awarded the Fields Medal in 2006 but declined it. In 2010 he declined $1 million also by the Clay Mathematics Institute for solving this million-dollar problem.

Riemann Hypothesis

Every zeta function counts something. Recall that in Chapter 4, it counted the number of solutions of equations over finite fields. To be precise, let N_p denote the number of solutions of the equation

$$y^2 = x^3 + ax + b \tag{15.1}$$

defining an elliptic curve E over $\mathbb{F}_p$ $(p > 2)$. To count the number of solutions of (15.1) we plug $x \neq 0$ from $\mathbb{F}_p$ in it. For half of them $f(x) = x^3 + ax + b$ is expected to be a square in $\mathbb{F}_p$, whereas for the other half, a non-square in $\mathbb{F}_p$. When $f(x) \neq 0$ is a square, there are two $y = \pm\sqrt{f(x)}$ that satisfy (15.1). If $f(x) = 0$, there is only one such y. Thus, the expected value of N_p is around

$$2 \cdot \frac{p-1}{2} + 1 = p.$$

The Riemann hypothesis for the zeta function of E gives a bound for the deviation $a_p = a_p(E) = p + 1 - N_p$ from the expected value $p + 1$ (counting the point at infinity on E) of N_p:

$$|a_p| = |p + 1 - N_p| \leq 2\sqrt{p}.$$

Note that the error a_p is of a lower order of magnitude than N_p.

Recall that (Chapter 13) Euler had already used the zeta function $\zeta(s)$, for real $\sigma > 1$ and $t = 0$ in $s = \sigma + it$, defined by the infinite series

$$\zeta(s) = \sum_{n=1}^{\infty} \frac{1}{n^s} \tag{15.2}$$

to prove the infinitude of prime numbers by noticing that $\lim_{\sigma \to 1+} \zeta(\sigma) = \infty$. Moreover, he calculated $\zeta(\sigma)$ for even $\sigma = 2, 4, 6, \ldots$.

To study the distribution of primes, the number of primes $p \leq x$ is denoted by $\pi(x)$. Two functions $f(x)$, $g(x)$ increasing monotonically to infinity are by definition of the *same order* if $\lim_{x \to \infty} \frac{f(x)}{g(x)} = 1$ (and denoted by $f \sim g$).

In 1798, Legendre conjectured that $\pi(x) \sim x/\log x$, $\log x$ being the natural logarithm. Gauss claimed that when he was 16, he had also conjectured that

$$\pi(x) \sim Li(x) = \int_2^x \frac{dt}{\log t}$$

(see Table 15.1).

TABLE 15.1: Gauss's table of $\pi(x)$ and $Li(x)$.

x	$\pi(x)$	$Li(x)$	Difference
500,000	41,556	41,606.4	50.4
1,000,000	78,501	78,627.5	151.1
1,500,000	114,112	114,263.1	126.5
2,000,000	148,883	149,054.8	171.8
2,500,000	183,016	183,245.0	229.0
3,000,000	216,745	216,970.6	225.6

Since $Li(x) \sim x/\log x$, they both had conjectured the **Prime Number Theorem** (PNT):

$$\boxed{\pi(x) \sim x/\log x.}$$

Almost a century after it was conjectured, the PNT was proved in 1896 independently by **Jaques Hadamard** (1865–1963) and **Charles de la Valée Pousin** (1866–1962).

If we write

$$\pi(x) = x/\log x + E(x),$$

where $E(x)$ is the error term, the PNT as stated above contains no information about $E(x)$. In order to study $E(x)$, Riemann [Rie] in 1859 extended $\zeta(\sigma)$ for all $s = \sigma + it$ in the complex plane, except for $s = 1$ where the series in (15.2) diverges, by the functional equation

$$\zeta(s) = 2^s \pi^{s-1} \sin\left(\frac{\pi s}{2}\right) \Gamma(1-s)\zeta(1-s), \tag{15.3}$$

$\Gamma(s)$ being the Euler's gamma function (see Chapter 13). We have $\zeta(s) = 0$ for $s = -2, -4, -6, \ldots$, called the *trivial zeros* of the Riemann zeta function $\zeta(s)$ defined by (15.2) and (15.3).

To get a good bound on $E(x)$, Riemann assumed the following, called the

Riemann Hypothesis. *The non-trivial zeros of $\zeta(s)$ all lie on the line $\sigma = \frac{1}{2}$ in the complex plane.*

Replacing $Li(x)$ with the Cauchy principal value of the divergent integral

$$\ell i(x) = \int_0^x \frac{dt}{\log t}$$

the Riemann hypothesis gives the "best possible" bound for the error term $E(x)$, namely

$$|\pi(x) - \ell i(x)| < \frac{\sqrt{x}\log x}{8\pi}$$

for all $x \geq 2657$.

Of course, the value $\pi(2657)$ is a general knowledge.

Poincaré Conjecture

It was explained in Chapter 13 what is meant by saying two topological spaces are homeomorphic. A topological space X is a *real n-dimensional manifold* or simply an *n-manifold* if each x in X has a neighborhood homeomorphic to an open ball in $\mathbb{R}^n$ of a positive radius centered at x. Moreover, some compatibility conditions are assumed on the intersections of intersecting neighborhoods. An n-manifold X is compact if it is bounded and contains its boundary. (Note that if ∂X denotes the boundary of X, then $\operatorname{diam} \partial X < \dim X$.) Every torus $\mathbb{T}_g$ $(g \geq 0)$ is a compact 2-manifold without boundary. Recall that the sphere $S^2 : x^2 + y^2 + z^2 = 1$ is $\mathbb{T}_0$. The unit circle $S^1 : x^2 + y^2 = 1$ in the plane is a compact 1-manifold, also without boundary. The open ball in the Euclidean 3-space is a 3-manifold. Its boundary is its surface, which is excluded, hence it is not compact.

The *fundamental group* $\pi(X, x_0)$ of a topological space X with base point x_0 is the set of all loops α in X starting and ending at x_0. If a loop α_1 can be pushed around, shrunk, or stretched to another loop α_2, then α_1 and α_2 are the same. The sum $\alpha + \beta$ of two loops α and β is α followed by β. This makes sense because α, β both start and end at x_0. If $\alpha + \beta$ is the same as $\beta + \alpha$ for all α, β in $\pi(X, x_0)$, then $\pi(X, x_0)$ is *abelian*, otherwise, non-abelian. If there is a path in X between every two points of X, the choice of the base point x_0 does not matter, and we write $\pi(X, x_0)$ as $\pi(X)$. Moreover, if X and Y are homeomorphic, they have the same fundamental group.

Since every loop on the sphere S^2 can be shrunk to a point, $\pi(S^2)$ is trivial. On the other hand, there is a basic loop α on S^1 – going once around it in the positive direction, which cannot be shrunk to a point. If β is any loop in S^1, then $\beta = n\alpha$, i.e., going around S^1 n-times, in the positive or negative direction depending on whether n is positive or negative. Thus, $\pi(S^1) = \mathbb{Z} = \{0, \pm 1, \pm 2, \ldots\}$. It is easy to see that $\pi(\mathbb{T}_1)$ is abelian, generated by two basic loops α, β whereas $\pi(\mathbb{T}_2)$ is non-abelian, generated by four basic loops $\alpha_1, \beta_1, \alpha_2, \beta_2$ with only one relation $\alpha_1 \beta_1 \alpha_1^{-1} \beta_1^{-1} \alpha_2 \beta_2 \alpha_2^{-1} \beta_2^{-1} = 1$ (see Figure 15.1.)

The Poincaré conjecture asserts the following:

- *A compact connected 3-manifold is homeomorphic to a 3-sphere S^3 if and only if its fundamental group is trivial.*

Here the 3-*sphere* S^3 is the set of all points (x_1, x_2, x_3, x_4) in $\mathbb{R}^4$ with $x_1^2 + x_2^2 + x_3^2 + x_4^2 = 1$.

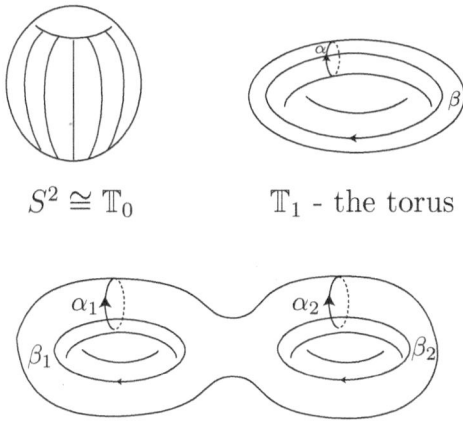

$S^2 \cong \mathbb{T}_0$ $\mathbb{T}_1$ - the torus

$\mathbb{T}_2$ Torus with two handles.

FIGURE 15.1: Tori.

Birch & Swinnerton-Dyer (B & S-D) Conjecture

Suppose E is an elliptic curve defined over $\mathbb{Q}$ by

$$y^2 = x^3 + ax + b. \tag{15.4}$$

We may assume that a, b are in $\mathbb{Z}$. The set $E(\mathbb{Q})$ of rational points on E is an abelian group under addition. (Recall from Chapter 10, how to add points of $E(\mathbb{Q})$.) The most celebrated theorem about $E(\mathbb{Q})$ is the following fact.

Mordell-Weil Theorem. *The group $E(\mathbb{Q})$ is finitely generated.*

This means that there are a finite number of points $P_1, \cdots, P_r$ in $E(\mathbb{Q})$ such that the points of $E(\mathbb{Q})$ are obtained by adding $P_1, \cdots, P_r$ in all possible ways. The maximum number r of points $P_1, \ldots, P_r$ linearly independent over $\mathbb{Z}$ (that is $m_1 P_1 + \cdots + m_r P_r = O \Rightarrow m_1 = \cdots = m_r = 0$) is called the *Mordell-Weil rank* or the *arithmetic rank* of E and written as $\mathrm{rk}(E)$. The determination of $\mathrm{rk}(E)$, and how large $\mathrm{rk}(E)$ can be, are some of the most challenging problems in the arithmetic of elliptic curves or more generally in what we call *arithmetic algebraic geometry*. One of the suggested tools to answer such questions is the Birch & Swinnerton-Dyer (B&S-D) conjecture that we explain now.

The Riemann zeta function is a product of its Euler factors

$$\zeta(s) = \prod_p \left(1 - \frac{1}{p^s}\right)^{-1},$$

the product being over all primes.

In a similar manner there is a so-called L-function $L_E(s)$ attached to the elliptic curve E defined by (15.4) as follows: Since (15.4) defines an elliptic curve, $\Delta = -16(4a^3 + 27b^2) \neq 0$. The reduction mod p ($p > 2$) of (15.4) defines an elliptic curve $E/\mathbb{F}_p$, provided p is a *good prime*, i.e., does not divide Δ. The factors in the Euler product of $L_E(s)$, $s = \sigma + it$ in $\mathbb{C}$, for good prime p are

$$L_p(s) = (1 - a_p p^{-s} + p^{1-2s})^{-1},$$

the reciprocals of the numerators of the zeta functions $Z_E(s)$ of $E/\mathbb{F}_p$ as p varies. (Note that the denominators of $Z_E(s)$ are independent of E.) Finally,

$$L_E(s) = \prod_p L_p(s), \tag{15.5}$$

the product taken over all good primes p. By the Hasse inequality:

$$|a_p| \leq 2\sqrt{p},$$

the infinite product on the right of (15.5) converges for $\sigma > \frac{3}{2}$, and its analytic continuation has, at $s = 1$, the Taylor expansion

$$L_E(s) = a_g(s-1)^g + a_{g+1}(s-1)^{g+1} + \cdots \tag{15.6}$$

with $a_g \neq 0$. The *analytic rank* of E is the integer $g \geq 0$ in (15.6) and is denoted by $\mathrm{ord}_{s=1}L_E(s)$.

A priori, there is no relation between the arithmetic $\mathrm{rk}(E)$ and the analytic rank $\mathrm{ord}_{s=1}L_E(s)$. The **Birch & Swinnerton-Dyer** conjecture of 1958 asserts that for every $E/\mathbb{Q}$,

- $\mathrm{rk}(E) = \mathrm{ord}_{s=1}L_E(s)$.

The weak version of the B & S-D conjecture states that

- $\mathrm{rk}(E) > 0 \Leftrightarrow \mathrm{ord}_{s=1}L_E(s) > 0$.

Birch & Swinnerton-Dyer formulated this conjecture by looking at the function

$$f(x) = \prod_{p \leq x} \frac{p}{N_p}.$$

Note that if we formally evaluate $L_E(s)$ at $s = 1$, we have

$$L_E(1) = \prod_{p \nmid 2\Delta} \frac{p}{N_p}.$$

Applications of B & S-D Conjecture

As early as the 10th century, the Persian mathematician al Karaji raised the question: which positive rationals a are areas of right triangles with all side lengths rational (see Chapter 10), equivalently, which of the square-free numbers

$$1, 2, 3, 5, 6, 7, 10, \ldots$$

are congruent numbers? In 1983, assuming the weak B & S-D conjecture, Tunnell published a complete solution of the above problem about congruent numbers.

(Tunnell). *Let $n_1(a)$ denote the number of triplets (x, y, z) of non-negative integers such that*

$$\begin{cases} a = 2x^2 + y^2 + 32z^2 \ if \ a \ is \ odd; \\ a/2 = 4x^2 + y^2 + 32z^2 \ if \ a \ is \ even. \end{cases}$$

Similarly, let $n_2(a)$ denote the number of (x, y, z) with

$$\begin{cases} a = 2x^2 + y^2 + 8z^2, \ if \ a \ is \ odd; \\ a/2 = 4x^2 + y^2 + 8z^2 \ if \ a \ is \ even. \end{cases}$$

Assuming the weak B & S-D Conjecture, a is a congruent number if and only if

$$n_2(a) = 2n_1(a).$$

Fermat's Last Theorem

The persistent attempts to solve such a simple Diophantine equation as

$$x^n + y^n = z^n \ (n \geq 3) \tag{15.7}$$

led to great advances in number theory, in particular, the creation of algebraic number theory. It is easy to see that it suffices to show that (15) has no non-trivial solution for prime values ℓ of n.

Fermat's Last Theorem *If $\ell \geq 3$ is a prime, then*

$$x^\ell + y^\ell = z^\ell \tag{15.8}$$

has no solution in integers with $xyz \neq 0$.

In 1847, **Ernst Kummer** (1810–1893) proved Fermat's Last Theorem (FLT) for the so-called regular primes. There are only three primes 37, 59,

67 less than 100, which are irregular. However, it is not known if there are infinitely many regular primes.

In 1984 at the Mathematical Research Institute, Oberwolfach, Germany, the German mathematician Gerhard Frey proposed a brilliant strategy to prove FLT (see [Fre]). This is based on a conjecture of Yutaka Taniyama and Goro Shimura and an observation of Hellegouarch. Suppose (15.8) has a non-trivial solution (a, b, c), that is

$$a^\ell + b^\ell = c^\ell \tag{15.9}$$

with $abc \neq 0$. Without loss of generality, we may assume that $\ell \geq 5$, $a \equiv -1 \pmod 4$ and b a multiple of 32. Look at the cubic equation

$$y^2 = x(x - a^\ell)(x + b^\ell). \tag{15.10}$$

The discriminant of the cubic (see Cardano's solution of the cubic, Chapter 11) can also be written, up to a constant, as

$$\Delta = ((\alpha_1 - \alpha_2)(\alpha_1 - \alpha_3)(\alpha_2 - \alpha_3))^2, \tag{15.11}$$

where $\alpha_1, \alpha_2, \alpha_3$ are its three roots. An *elliptic curve* E is defined by a Diophantine equation

$$y^2 = f(x) \tag{15.12}$$

where the three roots $\alpha_1, \alpha_2, \alpha_3$ of the cubic $f(x)$ are distinct. The quantity Δ defined by (15.11) is also called the discriminant of E. It turns out that Δ is a non-zero integer. Unfortunately, Δ depends on the Diophantine equations defining different but isomorphic elliptic curves. Since E stands for any of the isomorphic elliptic curves, $\Delta = \Delta_E$ needs to be taken the discriminant of, in some sense, a minimal equation (in the generalized Weierstrass form) defining E.) We can rewrite (15.12) as

$$y^2 = (x - \alpha_1)(x - \alpha_2)(x - \alpha_3).$$

For the *Frey curve* (15.10) we may assume that a, b, c have no common factors which would imply that $\alpha_1 = 0, \alpha_2 = a^\ell, \alpha_3 = -b^\ell$ are all distinct. Hence ((15.10)) defines an elliptic curve whose discriminant, up to a fixed power of 2, is given by

$$\Delta = (abc)^{2\ell}. \tag{15.13}$$

There is another invariant $N = N_E$ of the elliptic curve E called the *conductor*. The number N_E is roughly the product of distinct primes dividing Δ_E.

By the way, in view of (15.13) FLT for large ℓ would also follow at once from **Szpiro's conjecture**.

Conjecture (Szpiro). *Given $\epsilon > 0$, there is an absolute constant $c = c(\epsilon) > 0$ such that for all elliptic curves E,*

$$\Delta_E \leq c \cdot N_E^{6+\epsilon}.$$

Taniyama-Shimura Conjecture

We have seen in Chapter 7, the circle

$$x^2 + y^2 = 1$$

can be parameterized by rational functions

$$x(t) = \frac{t^2 - 1}{t^2 + 1}, \ y(t) = \frac{2t}{t^2 + 1}.$$

An elliptic curve E defined over $\mathbb{Q}$, or for short $E/\mathbb{Q}$ is *modular* if it is parameterized by modular functions. (See [Ser] for what modular functions are.) The Taniyama-Shimura conjecture (now a theorem) asserts that *every $E/\mathbb{Q}$ is modular*. Some people like to call it Taniyama-Shimura-Weil conjecture because it was A. Weil who made it widely known to the mathematical community.

Proof of FLT

Following the program laid out by Frey, FLT was proved in two steps by Ken Ribet and Andrew Wiles. Of course, there are numerous others whose work and ideas have been used by them in the proof. In a 1986 preprint, which was published in 1990, Ribet [Rib] showed that if a Frey curve (arising from a non-trivial solution of a Fermat equation) exists, then the Taniyama-Shimura conjecture is false for this elliptic curve.

(Ribet 1986). *The Taniyama-Shimura conjecture implies* FLT.

If a prime p divides the discriminant Δ_E of the elliptic curve E defined by the Diophantine equation (15.12), the reduction of $f(x) \bmod p$ has a repeated root α. Roughly speaking, E is *semistable* if for all such p, α has multiplicity two, not three. (The primes $p = 2, 3$ need special care.]

Actually, since Frey curves are semistable, one only needs to consider the case of the Taniyama-Shimura conjecture for semistable curves. On June 23, 1993 Andrew Wiles stunned the mathematical world by announcing a proof of the remaining part of this program.

"Theorem" (Wiles 1993). The Taniyama-Shimura conjecture holds for semistable curves.

There remained a gap in his proof of the "Theorem" that he finally filled in [Wil] (in collaboration with his former student Richard Taylor) in 1995 [T-W], completing the proof of FLT.

According to Sir Andrew Wiles, [Wil] is the proof of FLT, but according to others who also contributed to the proof, [Rib], [T-W], and [Wil] taken together is the complete proof of FLT.

Public Key Cryptography

The cryptosystems discussed in Chapter 4 are clandestine in the sense that the ciphering and deciphering keys of the cryptosystem are known only to the two parties involved in the transmission of the message. The ciphering key σ and its inverse, the deciphering key σ^{-1}, are symmetric in the sense that each of σ and σ^{-1} can be computed in about the same amount of time from knowing the other.

The public key is a cryptosystem where the ciphering key σ is public knowledge. In other words, it is published in a directory, and every user in the system can send a message to another user even without knowing this person. However, the success of any public key rests on the fact that σ^{-1} cannot be computed easily without some further information that is known only to the sender. More precisely, the **Public key** is a directory consisting of the following:

(i)   An alphabet $\aleph$, consisting of letters, digits, punctuation signs, etc., called characters.

(ii) For each user A, a key, that is a permutation $\sigma_A : \aleph \to \aleph$ such that σ_A^{-1} cannot be computed from the knowledge of σ_A. However, each A knows their σ_A^{-1} and keeps it confidential.

Suppose A wants to send a message (string of characters)

$$s = x_1 \cdots x_r$$

to B. To do so, A computes $\sigma = \sigma_B \sigma_A^{-1}$ and sends the messages as a string

$$s' = \sigma(x_1) \cdots \sigma(x_r) = y_1 \cdots y_r.$$

Since B knows both σ_B^{-1} and σ_A, the message can be recovered for

$$\sigma_A \sigma_B^{-1}(y_j) = \sigma_A \sigma_B^{-1}\left(\sigma_B \sigma_A^{-1}(x_j)\right) = x_j.$$

A public key called the **RSA** was invented in 1978 by Rivest, Shamir and Adleman. To explain it, we recall,

- **Euler's Theorem.** *If $n > 2$ is any integer and $1 \le a < n$ with* $\mathrm{GCD}(a, n) = 1$, *then* $a^{\phi(n)} \equiv 1 \pmod{n}$.

Here $\phi(n)$ is the Euler ϕ-function. Now suppose that for positive integers d and e,

$$de \equiv 1 \pmod{\phi(n)}$$

so that

$$a^{de-1} \equiv 1 \pmod{n}$$

or

$$a^{de} \equiv a \pmod{n}. \tag{15.14}$$

The congruence (15.14) is the main idea behind RSA.

In RSA every user A chooses two large primes p_A, q_A each with, say, 300 digits and keeps them confidential. Each A

  (i)   publishes $n_A (= p_A q_A)$ without revealing the factors p_A and q_A of n_A;

  (ii) chooses at random integer e_A coprime to $\phi(n_A)$ and publishes this e_A as well.

So, all in all, in the directory the key for each A is published as (n_A, e_A). Each A computes $d_A = e_A^{-1} \bmod \phi(n_A)$, so that

$$d_A e_A \equiv 1 \pmod{\phi(n_A)}.$$

However, A does not divulge d_A.

The alphabet $\aleph$ can be imbedded in $\mathbb{Z}/n_A\mathbb{Z} = \{\bar{0}, \bar{1}, \cdots, \overline{n_A - 1}\}$ as, say, the first $|\aleph|$ invertible elements of $\mathbb{Z}/n_A\mathbb{Z}$ and the arithmetic is done mod n_A. To send a message $s = a_1 \cdots a_r$ to B, what A sends is

$$s' = a_1^{e_B} \cdots a_r^{e_B} = b_1 \cdots b_r, \text{ say.}$$

Only B can recover the message s as

$$b_1^{d_B} \cdots b_r^{d_B} = a_1^{e_B d_B} \cdots a_r^{e_B d_B} = a_1 \cdots a_r = s.$$

for only B knows $d_B = e_B^{-1} \pmod{\phi(n_B)}$.

To break the public key (n_A, e_A) is to find the inverse $d_A = e_A^{-1}$ of $e_A \bmod \phi(n_A)$. To be able to do so one needs to know what $\phi(n_A)$ is. But see Chapter 14.

$$\phi(n_A) = n_A \left(1 - \frac{1}{p_A}\right)\left(1 - \frac{1}{q_A}\right).$$

Thus if one knows the factorization $n_A = p_A q_A$, which only A knows, one knows $\phi(n_A)$, and hence can decrypt. It may be remarked that conversely, if one knows n and $\phi(n)$, one can factor n in polynomial time. Thus for practical purposes, knowing $\phi(n)$ and knowing the factorization of n may be considered to be equivalent. It is an exceedingly difficult problem even for the fastest computers to factor such a large number into the (unique) product of primes. This ensures that it would be extremely difficult to break the code (n_A, e_A). For more, see [Kob].

Miscellaneous

There were some remarkable events during the 20th century that greatly influenced the development of mathematics. For example, the establishment of mathematical societies and journals throughout the world made it possible for mathematicians from all over the world to communicate with each other. A whole chapter could be devoted to the rise of Japan as a world power in mathematics, or to some unique institutions that can only be identified with meetings among like-minded mathematicians to work on specific projects. One could also write about some unusual personalities like Ramanujan, **Emmy Noether** (1882–1935) one of the greatest (female) mathematicians of all time, or more recently **Paul Erdös** (1913–1996), a wandering mathematician and a saintly person with no interest in worldly possessions. Perhaps the most revolutionary mathematician of the 20th century is Alexander Grothendieck. He, and his school, did to the 20th century mathematics what Cantor did to the mathematics of the 19th century. The following are only selections.

Bourbaki

Just before the 2nd World War, a group of French mathematicians banded together under the fictitious name of Nicolas Bourbaki to write (like Euclid's *Elements*) a series of books called the *Elements of Mathematics* with the aim of giving a systematic and rigorous exposition of current mathematics. Each book, called a chapter, is devoted to one topic. Some of them, e.g. *Commutative Algebra*, have become indispensable for researchers in their fields. Among Bourbaki's founding members are H. Cartan, C. Chevalley, J. Dieudonné and A. Weil. The first Bourbaki Conference was held in July 1936. For details, see [Wei-2]. The abstraction initiated by Cantor and completed by Grothendieck and others (of the Bourbaki school) became very fashionable and peaked during the third quarter of the 20th century. In American schools, the teaching of set theory under the banner of "new math" became popular. However, it proved disastrous for the lack of teachers adequately trained in conceptual mathematics. The following anecdote would not be out of context.

In a school in Baltimore, Maryland, the pupils were being drilled about the qualifier notation in set theory. The teacher made all the students in the class whose fathers were doctors stand up and said, "this is the set of students in the class whose fathers are doctors." Then those students stood up who did not do their homework and the teacher said, "this is the set of students who did not do their homework." Then those had to stand up who ..., and the drill went on. At the end of the class period, a professor of mathematics who was visiting the class (his son was one of the pupils) asked the class, "is a carton of eggs a set?" All the students shouted, "No!". "Why not?", asked the professor. The students replied with one voice, "because the eggs cannot stand up."

Fields Medal

The Fields Medal is the most prestigious prize in mathematics. At the time it is awarded, the recipient must be under 40 years of age. Besides, these medals are awarded only once every 4 years. J.-P. Serre is the youngest recipient of a Fields Medal so far, which he received in 1954 at the age of 28. He made outstanding contributions to algebra, algebraic geometry, number theory and topology. The discussion of his work is well beyond the scope of this book. He has also written several books with unrivaled clarity of exposition and made a number of conjectures. These conjectures have greatly influenced development of mathematics during the second half of the 20th century. For example, Ribet's Theorem, which is the first step in the proof of Fermat's Last Theorem is a consequence of a conjecture of Serre.

There are some peculiarities with the Fields Medal. As mentioned earlier, the recipient must be under 40 years old at the time of the award. If Sir Andrew Wiles did not have a gap in his proof of Fermat's Last Theorem, he would certainly be a Fields Medalist. By the time he fixed the gap, he was barely over 40. Besides, there are mathematicians who are of higher order than a Fields Medalist, but somehow missed the Fields Medal. Among others, perhaps for this reason also, the Abel Prize in mathematics, named after the Norwegian mathematician Niels Abel, was instituted by the Norwegian Academy of Sciences and Letters. Another almost equally prestigious prize for mathematics is the Wolf Prize awarded by the Wolf Foundation in Israel.

All Fields Medalists and Abel Laureates are listed at the end of the book.

Hardy and Ramanujan

In the history of mathematics, the story of S. Ramanujan is unique. He was born in Erode, near Madras (now Chennai), India. In 1903 he passed the matriculation examination (to obtain a high school certificate) and won a scholarship to enter a college for further studies. However, at college, he devoted his entire time to mathematics and neglected other subjects, and failed in those subjects. As a result, his scholarship was revoked and he had to drop out of college. In 1906 he took the job of clerk at the Port Trust in Madras, devoting most of his free time to mathematics. He was a mathematical genius of the highest order, but with no formal training in mathematics. He was a self-taught mathematician using Carr's *Synopsis of Pure Mathematics*, an unusual book meant to help English students pass examinations. It contains thousands of theorems, but very few with proofs. Ramanujan took it upon himself to prove all the formulas in the book. At the urging of influential friends, on January 16, 1913 Ramanujan sent some of his papers and introduced himself to G.H. Hardy, a leading professor of pure mathematics at Cambridge University. Ramanujan's genius was immediately recognized and Hardy made arrangements to bring Ramanujan to Cambridge. This resulted in one of the most fruitful collaborations in the history of mathematics. The change in the environment did not suit Ramanujan. The weather in England was very different from that in his native Madras. Being a vegetarian, he had to cook for himself which he often neglected to do, for he was too absorbed in mathematics. He soon fell ill, and died in 1920. He left behind a number of published and unpublished works that are still a source of inspiration for mathematicians to work on. It is a challenge even for world-class mathematicians to provide missing proofs to fully understand his discoveries.

To know the depth of his work, just consider the *Ramanujan tau function* $\tau(n)$ defined by the identity:

$$q \prod_{j=1}^{\infty} (1 - q^j)^{24} = \sum_{n=1}^{\infty} \tau(n) q^n.$$

The exponent 24 suggests that $\tau(n)$ has something to do with expressing n as a sum of 24 squares.

In 1916, Ramanujan discovered, but did not prove, that

- $\tau(n)$ *is multiplicative*, i.e., $\tau(mn) = \tau(m)\tau(n)$ if GCD $(m, n) = 1$.

- *For p prime and $r > 0$,*

$$\tau(p^{r+1}) = \tau(p)\tau(p^r) - p^{11}\tau(p^{r-1}).$$

- $|\tau(p)| \leq 2p^{11/2}$.

The first two of the above three properties of $\tau(n)$ were proved a year later in 1917 by Mordell.

Note the similarity of the third one, the so-called *Ramanujan conjecture*, to the Weil inequality (4.11) discussed in Chapter 4. The Ramanujan conjecture was proved by Pierre Deligne in 1974 as a consequence of his proof of the Weil conjecture for which he was awarded the Fields Medal.

The similarity becomes more apparent when one considers the Ramanujan L-function defined for $s = \sigma + it$ either by

$$L(s) = \sum_{n=1}^{\infty} \frac{\tau(n)}{n^s}$$

for $\sigma > 6$ or its continuation via the fundamental equation

$$\frac{L(s)\Gamma(s)}{(2\pi)^s} = \frac{L(12-s)\Gamma(12-s)}{(2\pi)^{12-s}}.$$

For $\sigma > 7$, it has the Euler product:

$$L(s) = \prod_{p \text{ prime}} \frac{1}{1 - \tau(p)p^s - p^{11-2s}}.$$

Ramanujan conjectured that all non-trivial zeros of $L(s)$ are on the line $\sigma = 6$, equivalently

$$|\tau(p)| \le 2p^{11/2}.$$

By the way, Lehmer conjectured that $\tau(n)$ is never zero. In 2013, van Hoeij and Zen checked Lehmer's conjecture for all $n \le 861\,212\,624\,008\,487\,344\,127\,999$, but the conjecture remains unproved.

In number theory, the partition function $p(n)$ is the number of ways to partition n as a sum of positive integers. For example, $p(4) = 5$ as 4 is the sum $1+1+1+1, 1+1+2, 1+3, 2+2$, and 5. The value of $p(n)$ for $n = 1, 2, 3, 4, 5, 6, \ldots$ is $1, 2, 3, 5, 7, 11, \ldots$, respectively. It grows very fast, through, e.g., $p(100) = 190,569,292$. There is no closed formula for $p(n)$. However, Euler found the following equation to find $p(n)$ recursively:

$$\sum_{n=0}^{\infty} p(n)x^n = \prod_{k=1}^{\infty} (1-x^k)^{-1}.$$

In 1918, Hardy and Ramanujan proved that

$$p(n) \sim \frac{1}{4n\sqrt{3}} \exp\left(\pi\sqrt{\frac{2n}{3}}\right).$$

(Recall that $f(x) \sim g(x)$ means $\lim_{x\to\infty} \frac{f(x)}{g(x)} = 1$ and $\exp(x) = e^x$.)

Ramanujan also discovered that

$$p(5k+4) \equiv 0 \pmod{5}$$
$$p(7k+5) \equiv 0 \pmod{7}$$
$$p(11+6) \equiv 0 \pmod{11}$$

and no such congruence holds for a prime $\ell \ne 5, 7$ and 11.

That a first- or second-year college dropout could discover such a mathematics, which took the combined efforts of several Fields Medalists to prove, is no less than divine. In fact, Ramanujan claimed that he received his mathematics in dreams from the goddess Namagiri.

Many consider Hilbert to be the greatest mathematician of the 20th century. Besides his own contributions, he also gave a final touch to the mathematics that was known before his time.

When Hardy was asked to rate some of the mathematicians of his time on a scale of 1–100, he gave himself the score 25, Hilbert 80, and Ramanujan 100.

Mathematics in America

Today America is a leader in mathematics, but this is a relatively recent phenomenon. Until around the beginning of the last quarter of the 19th century, there was hardly any mathematics worth mentioning in the United States. **J. J. Sylvester** (1814–1897) was the first mathematician of repute to come to America in 1876 as a professor of mathematics at the newly opened Johns Hopkins University, the first research-oriented university in the United States. In a radical departure from other universities, it was primarily a graduate school. At Johns Hopkins, Sylvester in 1878 founded the *American Journal of Mathematics*, the first mathematical journal in America, and helped develop a tradition of graduate studies in mathematics in the United States. Johns Hopkins University invited such outstanding mathematicians as **O. Zariski** (1899–1986), **W.-L. Chow** (1911–1995) and **K. Kodaira** (1915–1997) to join its faculty. The University of Chicago is another institution that can boast of such legendary figures as **L.E. Dickson** (1874–1954), **A. Weil** (1906–1998), and **S.S. Chern** (1911–2004) who served on its faculty. In 1928 the University of Chicago awarded about one third of all the Ph.D.s in mathematics in the United States. It has produced such outstanding mathematicians as G.D. Birkoff, L.E. Dickson and R.L. Moore.

Another notable mathematician whose work has already been mentioned in Chapter 4 is Emil Artin, an Austrian who emigrated to the United Sates in 1937. His longest tenure was at Princeton University. He was one of the leading mathematicians of the 20th century. He supervised Ph.D. theses of the next generation of several prominent American mathematicians such as **John T. Tate** (1925–2019). Tate himself supervised the Ph.D. theses of several eminent mathematicians. The work of two of them (Ribet and Tunnell) has already been mentioned in this book.

Once I asked Tate how did he produce some of the best mathematicians in America, and his short answer was, "I listen to them." Now the United States has some of the most elite centers of mathematics in the world, such as MIT, Princeton, Harvard and the University of California at Berkeley, to name a few. Often, some of the prominent faculty members are visitors from abroad, such as Gerd Faltings at Princeton from 1984 to 1994.

In 1984, Faltings was awarded the Fields Medal for his proof of the Mordell conjecture. The **Mordell conjecture** asserts the following:

if X is a curve defined over $\mathbb{Q}$ and of genus > 1, then the set $X(\mathbb{Q})$ of points on X with rational coordinates is finite.

The conjecture was later generalized by replacing $\mathbb{Q}$ with any number field.

Roughly speaking, if X is a non-singular algebraic curve defined by, say, $f(x,y) = 0$ of degree $d > 0$, its genus $g = g(X) = \frac{(d-1)(d-2)}{2}$. For example, the elliptic curve defined by (15.4) has genus 1.

It was known at the beginning of the 20th century that if $g(X) = 0$ and has one rational point on it, then $X(\mathbb{Q})$ is infinite. (The curve $X : x^2 + y^2 + 1 = 0$ has genus zero but no rational points on it.) Thus Faltings' theorem reduces the problem whether or not a curve has infinitely many rational points to the genus 1 case, which are elliptic curves. An elliptic curve may or may not have infinitely many rational points. At the moment, there is no proven algorithm to decide whether a given elliptic curve has only finitely many rational points.

Fields Medal Recipients

1936

Lars Valerian Ahlfors

Jesse Douglas

1950

Laurent Schwartz

Atle Selberg

1954

Kunihiko Kodaira

Jean-Pierre Serre

1958

Klaus Friedrich Roth

René Thom

1962

Lars Hörmander

John Willard Milnor

1966

Michael Francis Atiyah

Paul Joseph Cohen

Alexander Grothendieck

Stephen Smale

1970

Alan Baker

Heisuke Hironaka

Serge Novikov

John Griggs Thompson

1974

Enrico Bombieri

David Bryant Mumford

1978

Pierre René Deligne

Charles Louis Fefferman

Gregori Alexandrovitch Margulis

Daniel G. Quillen

1982

Alain Connes

William P. Thurston

Shing-Tung Yau

1986

Simon K. Donaldson

Gerd Faltings

Michael H. Freedman

1990

Vladimir Drinfeld

Vaughan F.R. Jones

Shigefumi Mori

Edward Witten

1994

Jean Bourgain

Pierre-Louis Lions

Jean-Christophe Yoccoz

Efim Zelmanov

1998

Richard E. Borcherds

W. Timothy Gowers

Maxim Kontsevich

Curtis T. McMullen

Andrew J. Wiles (Special Tribute)

2002

Laurent Lafforgue

Vladimir Voevodsky

2006

Andrei Okounkov

Grigori Perelman[1]

Terence Tao

Wendelin Werner

2010

Stanislav Smirnov

Cédric Villani

2014

Artur Avila

Manjul Bhargava

Martin Hairer

Maryam Mirzakhani

2018

Caucher Birkar

Alessio Figalli

Peter Scholze

Akshay Venkatesh

2022

Hugh Duminil-Copin

June Huh

James Maynard

Maryna Viazovska

---

[1]Grigori Perelman declined to accept the Fields Medal.

Abel Prize Laureates

2003

Jean-Pierre Serre

2004

Michael Atiyah

Isadore Singer

2005

Peter Lax

2006

Lennart Carleson

2007

S.R. Srinivasa Varadhan

2008

Jacques Tits

John Griggs Thompson

2009

Mikhail Gromov

2010

John Tate

2011

John Willard Milnor

2012

Endre Szemerédi

2013

Pierre René Deligne

2014

Yakov Sinai

2015

John F. Nash, Jr.

Louis Nirenberg

2016

Andrew John Wiles

2017

Yves Meyer

2018

Robert P. Langlands

2019

Karen Uhlenbeck

2020

Hillel Furstenberg

Gregory Margulis

2021

László Lovász

Avi Wigderson

2022

Dennis Sullivan

2023

Luis A. Caffarelli

2024

Michel Talagrand

2025

Masaki Kashiwara

Bibliography

[Apo] T. M. Apostol & M. A. Mnatsakanian, *A fresh look at the method of Archimedes*, American Math. Monthly, **111** (2004), 496–508.

[Asc] A. Asher & M. Ascher, *Code of the Quipu*, Univ. of Michigan Press (1981).

[Bak-1] A. Baker, *Transcendental Number Theory*, Cambridge University Press (1975).

[Bak-2] R. C. Baker (Ed.), *Euler Revisited*, Kendrick Press, Heber City, UT (2006).

[Ber] J. L. Berggren, *Episodes in the Mathematics of Medieval Islam*, Springer Verlag (1986).

[Bom] E. Bombieri, *Counting points on curves over finite fields* (d'après S. A. Stepanov), Séminaire Bourbaki **430** (1972–73).

[Bro] F.E. Browder (Ed.), *Mathematical Developments arising from Hilbert's Problems*, AMS (1976).

[Caj] F. Cajori, *A History of Mathematical Notations*, 2 vols., Dover (1993).

[Cal] R. Calinger (Ed.), *Classics of Mathematics*, Prentice Hall (1995).

[Car] G. Cardano, *Ars Magna*, Dover (1968).

[Cha-1] J.S. Chahal, *Topics in Number Theory*, Plenum (1988).

[Cha-2] J.S. Chahal, *Congruent Numbers and elliptic curves*, Amer. Math. Monthly, **113** (2006) 308–317.

[Clo] M. P. Closs (Ed.), *Native American Mathematics*, Univ. of Texas Press (1986).

[Dat] B. Datta & A.N. Singh, *History of Hindu Mathematics*, A Source Book, Lohore (1935-38).

[Ded] R. Dedekind, *Essays on the Theory of Numbers*, Dover (1963).

[Dic] L. E. Dickson, *Fermat's Last Theorem and the origin and nature of the theory of algebraic numbers*, Ann. Math. **18** (1917), 161–187.

[Dun] W. Dunham, *Journey through Genius*, Penguin (1991).

[Euc] Euclid, *The Thirteen Books of the Elements*, (Translation and Commentary by Sir T. L. Heath) vol. I, II, III, Dover (1956).

[Fib-1] Fibonacci, *Liber Abaci*, Springer (2002).

[Fib-2] Fibonacci, *Liber Quadratorum* (1225).

[Fre] G. Frey, *Links between stable elliptic curves and certain diophanatine equations*, Ann. Univ. Saraviensis, **1** (1986), 1–40.

[Gau] C. F. Gauss, *Disquisitiones Arithmeticae*, Yale University Press (1966).

[Gil] R. J. Gillings, Mathematics at the Time of the Pharaohs, Dover (1982).

[Göd] K. Gödel, *The Consistency of the Continuum Hypothesis*, Princeton Univ. Press (1940).

[GPY] D. A. Goldston, J. Pinz, and C. Y. Yildirim, *The path to recent progress on small gaps between primes*, Clay Math. Proc., **7** (2007), 129–139.

[Gre] M. J. Greenberg, *Euclidean and Non-Euclidean Geometry, Development and History*, Freeman (1993).

[Har] R. Hartshorne, *Geometry: Euclid and Beyond*, Springer Verlag (2000).

[Hil-1] D. Hilbert, *Grundlagen der Geometrie*, Teubner (1930).

[Hil-2] D. Hilbert, *Foundation of Geometry*, Open Court Pub. Co. (1971).

[Hog-1] L. Hogben, *Mathematics for the Millions*, Norton & Co, 1st Ed. (1937).

[Hog-2] J.P. Hogendijk, *Ibn al-Haytham's Completion of the Conics*, Springer Verlag (1985).

[Hør] J. Hørup, *Lengths, Widths, Surfaces–A Portrait of Old Babylonian Algebra and its Kins*, Springer Verlag (2002).

[Ito] K. Itô (Ed.), *Encyclopedic Dictionary of Mathematics*, MIT Press (1986).

[Jon] A. Jones, *Pappus of Alexandria–Book 7 of the Collection*, Springer Verlag (1986).

[Jos] G. G. Joseph, *The Crest of the Peacock, Non-European Roots of Mathematics*, Princeton University Press (2000).

[Kle] F. Klein, *Famous Problems of Elementary Geometry*, Dover (1956).

[Kob] N. Koblitz, *A Course in Number Theory and Cryptography*, 2nd ed., Springer (1994).

[Mar] J.-C. Martzloff, *A History of Chinese Mathematics*, Springer Verlag (1997).

[Maz] B. Mazur, *Imagining Numbers*, Farrar, Strauss and Giroux (2003).

[Men] K. Menninger, *Number Words and Number Symbols*, Dover (1992).

[Mer] R. Merris, *Graph Theory*, John Wiley (2001).

[Nar] M. S. Narasimhan, et. al., *Algebraic Topology*, TIFR Math. Pamphlet No. 2 (1964).

[Rib] K. Ribet, *From Taniyama-Shimura conjecture to Fermat's Last Theorem*, Ann. Fac. Sci. Toulouse, **11** (1990), 116–139.

[Rie] B. Riemann, *Über die Anzahl der Primzahlen unter einer gegebenen Grösse*, Gesammelte, Math., Monats. Preuss. Akad. Wiss. (1859–1860), 671–680.

[Ser] J.-P. Serre, *A Course in Arithmetic*, GTM 7, Springer (1973).

[Sch-1] W. M. Schmidt, *Norm form equations*, Ann. Math. **96** (1972) 526–551.

[Sch-2] W. M. Schmidt, *Equations over Finite Fields: An Elementary Approach*, Kendrick Press (2004).

[Ses] J. Sesiano, *Books IV–VII of Diophantus' Arithmetica*, Springer Verlag (1982).

[T-W] R. Taylor & A. Wiles, *Ring-theoretic properties of certain Hecke algebras*, Ann. Math., **141** (1995), 553–572.

[Too] G. J. Toomer (Ed.), *Apollonius of Persia, Conics*, vols. I and III, Springer Verlag (1990).

[Tun] J. Tunnell, *A classical Diophantine problem and modular forms of weight 3/2*, Invent. Math. **72** (1983), 323–334

[Var] V.S. Varadarajan, *Algebra in Ancient and Modern Times*, Hindustan Book Agency (1997).

[Wae-1] B. L. van der Waerden, *Geometry and Algebra in Ancient Civilization*, Springer Verlag (1983).

[Wae-2] B. L. van der Waerden, *A History of Algebra: From al-Khworizmi to Emmy Noether*, Springer Verlag (1985).

[Wan]   M. L. Wantzel, *Recherches sur les moyens de reconnaitre si un Prob-
lème de Géométrie peut se résoudre avec la règal et le compas*, J. de
Mathematiques Pures et Appl., **2** (1837) 366–372.

[Wei-1] A. Weil, *Number-Theory–An Approach Through History*, Birkhäuser
(1983).

[Wei-2] A. Weil, *The apprenticeship of a mathematician*, Birkhäuser (1992).

[Wil]   A. Wiles, *Modular elliptic curves and Fermat's Last Theorem*, Ann.
Math. **142** (1995), 443–551.

[Zha]   Y. Zhang, *Bounded gaps between primes*, Ann. Math., **179** (2014),
819–882.

Index